CYCLING AND RECYCLING

The Environment in History: International Perspectives

Series Editors: Dolly Jørgensen, *University of Stavanger;* Christof Mauch, *LMU Munich;* Kieko Matteson, *University of Haiwai'i at Mānoa;* Helmuth Trischler, *Deutsches Museum, Munich*

Volume 1
Civilizing Nature: National Parks in Global Historical Perspective
Edited by Bernhard Gissibl, Sabine Höhler, and Patrick Kupper

Volume 2
Powerless Science? Science and Politics in a Toxic World
Edited by Soraya Boudia and Natalie Jas

Volume 3
Managing the Unknown: Essays on Environmental Ignorance
Edited by Frank Uekötter and Uwe Lübken

Volume 4
Creating Wildnerness: A Transnational History of the Swiss National Park
Patrick Kupper
Translated by Giselle Weiss

Volume 5
Rivers, Memory, and Nation-Building: A History of the Volga and Mississippi Rivers
Dorothy Zeisler-Vralsted

Volume 6
Fault Lines: Earthquakes and Urbanism in Modern Italy
Giacomo Parrinello

Volume 7
Cycling and Recycling: Histories of Sustainable Practices
Edited by Ruth Oldenziel and Helmuth Trischler

Cycling and Recycling
Histories of Sustainable Practices

Edited by
Ruth Oldenziel and Helmuth Trischler

First published in 2016 by
Berghahn Books
www.berghahnbooks.com

© 2016, 2019 Ruth Oldenziel and Helmuth Trischler
First paperback edition published in 2019

All rights reserved. Except for the quotation of short passages
for the purposes of criticism and review, no part of this book
may be reproduced in any form or by any means, electronic or
mechanical, including photocopying, recording, or any information
storage and retrieval system now known or to be invented,
without written permission of the publisher.

Library of Congress Cataloging-in-Publication Data

Names: Oldenziel, Ruth, 1958- editor. | Trischler, Helmuth, editor.
Title: Cycling and recycling : histories of sustainable practices / edited by Ruth Oldenziel and Helmuth Trischler.
Other titles: Environment in history ; v. 7.
Description: First edition. | New York : Berghahn Books, 2015. | Series: Environment in history : international perspectives ; . 7 | "This volume looks at the bicycle in tandem with the history of recycling, and how both have shaped and are shaping Earth's environment." | Includes bibliographical references.
Identifiers: LCCN 2015014068| ISBN 9781782389705 (hardback : alk. paper) | ISBN 9781782389712 (ebook)
Subjects: LCSH: Cycling—History. | Cycling—Environmental aspects. | Recycling (Waste, etc.)—History. | Recycling (Waste, etc.)—Environmental aspects. | Sustainability.
Classification: LCC HE5736 .C926 2015 | DDC 363.72/8209—dc23
LC record available at http://lccn.loc.gov/2015014068

British Library Cataloguing in Publication Data

A catalogue record for this book is available from the British Library

ISBN 978-1-78238-970-5 hardback
ISBN 978-1-78920-055-3 paperback
ISBN 978-1-78238-971-2 ebook

Contents

List of Figures — vii

How Old Technologies Became Sustainable: An Introduction — 1
Ruth Oldenziel and Helmuth Trischler

I. CYCLING HISTORIES

Chapter 1. Use and Cycling in West Africa — 15
Hans Peter Hahn

Chapter 2. The Politics of Bicycle Innovation: Comparing the American and Dutch Human-Powered Vehicle Movements, 1970s–Present — 33
Manuel Stoffers

Chapter 3. Scarcity, Poverty, Exclusion: Negative Associations of the Bicycle's Uses and Cultural History in France — 58
Catherine Bertho Lavenir

Chapter 4. Who Pays, Who Benefits? Bicycle Taxes as Policy Tool, 1890–2012 — 73
Adri Albert de la Bruhèze and Ruth Oldenziel

Chapter 5. Monuments of Unsustainability: Planning, Path Dependence, and Cycling in Stockholm — 101
Martin Emanuel

II. INTERSECTIONS

Chapter 6. Bicycling and Recycling in Japan: Divergent Trajectories — 125
M. William Steele

III. RECYCLING HISTORIES

Chapter 7. Premodern Sustainability? The Secondhand and Repair Trade in Urban Europe — 147
Georg Stöger

Chapter 8. Waste to Assets: How Household Waste Recycling Evolved
    in West Germany   168
        Roman Köster

Chapter 9. Ecological Modernization of Waste-Dependent
    Development? Hungary's 2010 Red Mud Disaster   183
        Zsuzsa Gille

Chapter 10. The Scramble for Digital Waste in Berlin   202
        Djahane Salehabadi

IV. REFLECTIONS

Chapter 11. Can History Offer Pathways to Sustainability?   215
        Donald Worster

Chapter 12. History, Sustainability, and Choice   219
        Robert Friedel

Select Bibliography   226

Index   240

List of Figures

1.1.	Bicycle transport of ceramics	23
2.1.	An HPV utopia: human-powered vehicles on the road, on the water, and in the air	35
4.1.	Bicycles' Share of the Total Number of Car, Public Transport, Bicycle, and Moped Trips in Nine European Cities, 1920–1995	80
4.2.	Dutch Cyclists' Contribution to Interwar Road Building	82
4.3.	Traffic Count by Zones (The Hague, 1937)	83
4.4.	Cyclists Sidelined	83
5.1.	"Trafikledsplan för Stockholm"	106
5.2.	"Fairy tale about the tiny, tiny walkway on the huge, huge bridge"	115
6.1.	Abandoned bicycles in Kichijōji	133
6.2.	Mitaka City, Disposal of Impounded Bicycles, 2000–2009	135
6.3.	Musashino City, Disposal of Impounded Bicycles, 2003–2007	135
7.1.	Licenses on the Viennese *Tandelmarkt*, 1772–1791	152
7.2.	Mean annual taxes of cobblers on the Viennese *Tandelmarkt*, 1772–1791	155
7.3.	Prices of Men's Clothing in Salzburg, 1770–1790	156
7.4.	Secondhand Traders Taxed by the Viennese Magistracies, 1738–1803	158
8.1.	Production of Plastics in West Germany	172
8.2.	Gross Production Value of West Germany's Packaging Industry	172
8.3.	Consumption of Paper and Cardboard in West Germany, 1950–1970	175
8.4.	Prices for Waste Paper (Average per Year)	175
8.5.	Return Rate for Scrap Glass	177

How Old Technologies Became Sustainable
An Introduction

Ruth Oldenziel and Helmuth Trischler

Why bring the story of cycling and recycling together in one frame to understand and analyze how history can help us move toward more sustainable societies? Do the histories of commuting by city bikes and recycling of used bottles have anything to do with each other in the transition to sustainability?

On the surface, the idea of combining the history of waste treatment and recycling with the history of cycling and mobility seems like a bold undertaking. Today's discussions about sustainable technologies tend to focus on finding new solutions to pressing environmental challenges. The belief and hope that technological innovations will offer an escape route from impending ecological collapse is as pervasive as it is appealing. The belief in "green tech," for example, promises to avoid back-to-nature traditions, which some environmental activists have embraced as sound and sustainable, but is ridiculed by their opponents as sentimental and untenable. In this volume, however, we examine alternative debates. Our *Re/cycling* concentrates on the notion of transitioning to a more sustainable future by resurrecting older technologies for a new purpose. We explore the intriguing histories of two technologies that were advanced almost fifty years ago as important tools for a more sustainable future: cycling and recycling. As we argue, the two technologies have more than merely etymological similarities.

From the traditional viewpoint of the history of technology, waste treatment and bicycle production seem to have little in common. When approaching the same subjects from the perspective of consumers of goods and users of technology, however, we find that they are interrelated—certainly in practice, if not in theory. In the late 1960s and early 1970s, environmental activists mobilized older rather than newer technologies as political tools to save the planet. At the time, the revival represented a deliberate act of resistance to the politics of economic growth. Consumer activists demanded that glass bottles be returned to manufacturers. Cargo bikes were appropriated as an alternative to automobiles. Windmills were invested with the hope that they would one day replace nuclear power plants. Once ridiculed as hopelessly outdated, old technologies were deliberately embraced: they were revived through repurposing them into new uses and having new meanings reassigned to them.

In resurrecting older technologies for a new purpose, the rebelling consumers and users of the 1970s were pivotal in a movement that quickly became transatlantic and transnational. Many environmentalists in Europe and beyond found inspiration in their counterparts in the United States.[1] The influential San Francisco Bay Area entrepreneur Stewart Brand and his Whole Earth network, for example, placed the greatest hope for the environment on technology users. Brand advocated a do-it-yourself culture and believed in the transformative power of relevant technology. *The Whole Earth Catalogue: Access to Tools*, the first edition of which was issued in 1968, was a kind of shopping catalogue for the environmental movement: from educational instruments such as books, maps, and courses to well-designed, special-purpose utensils such as garden tools, welding equipment, and hiking gear. It listed tools for a just and sustainable society available on the market and offered people access to these instruments.[2] The reader could "find his own inspiration, shape his own environment, and share his adventure with whoever is interested."[3] The catalogue's mission was based on the ethic of do-it-yourself crafting, tinkering, and self-reliance; low-tech and high-tech tools as well as old and new implements were all part of the same universe.[4]

The do-it-yourself practice was also part of a new theory of appropriate technology. Originally defined by economist Ernst Friedrich Schumacher in his book *Small is Beautiful: A Study of Economics As If People Mattered* (1973), appropriate technology is an ideological movement believing that technological choices and their applications should be small-scale, decentralized, labor intensive, energy efficient, environmentally sound, and locally controlled. In the same spirit, the nascent British environmental movement published do-it-yourself books like the *Consumers' Guide to the Protection of the Environment*, which teaches consumers how to organize recycling clubs. Another publication, the *Environmental Handbook*, suggested that consumers should mobilize the law and rely on "maintenance and repair of existing products" instead of buying into the consumer-society logic of "planned obsolescence."[5]

The belief in low-tech and repair practices has endured and recently revived. In California's Bay Area, the Maker Faire movement, founded by *Make* magazine in 2006, promotes environmental resilience through a low-tech, do-it-yourself culture. In Africa, events were organized in cities like Accra (Ghana), Nairobi (Kenya), and Cairo (Egypt) to embrace "arts, crafts, engineering, science projects, and the do-it-yourself mindset." We find similar ideas in the Repair Café Foundation, initiated in 2009 by the Dutch former journalist Martine Postma, who inspired like-minded activists to create their own Repair Cafés in many European and American cities. Other grassroots movements provide open access to technological platforms, like fab labs (fabrication laboratory) and the Open Source Ecology network. These green tech initiatives all utilize old and new methods to generate low-tech and low-carbon technologies that users can apply in order to serve their communities and the planet. They are also a testimony to the long-lasting effect of the 1970s movement. It begs the question, however: to what extent does the 1970s represent a break with the past?

Historicizing Sustainability

Contemporaries—and others since then—experienced the 1970s as ground zero for the planet, as a sudden and seismic rupture in history, as if everything that had happened before was merely relegated to history in the face of the awful future threatening mankind. The 1970s were culturally reframed as radically different from earlier decades. At the same time, the period witnessed the celebration and resurrection of older practices and technologies, suggesting continuities to rather than a radical break from the past. Indeed, recently there has been an interest in recovering older notions of sustainability.

In 2013, nearly every town in Germany staged a day, if not a week, of sustainability (*Nachhaltigkeit*) in celebration of the "Year of Sustainability," commemorating the three-hundredth anniversary of the publication of Hans Carl von Carlowitz's *Sylvicultura oeconomica or the Instructions for Wild Tree Cultivation* (*Sylvicultura oeconomica oder Haußwirthliche Nachricht und Naturmäßige Anweisung zur Wilden Baum-Zucht Anweisung zur wilden Baum-Zucht*). For example, in the small town of Püttlingen, in the country's western state of Saarland, events such as the designing of apiary-friendly gardens to fight the devastation of the dying bee colonies were held. German communities embraced the mining officer Carlowitz as the true inventor of the term *sustainability*—as Germany's gift to the current global debate[6]—although he may seem like an unlikely hero for today's environmental challenges. Before *Sylvicultura oeconomica* was published in 1713, he had been managing mines on behalf of the Saxon court in Freiberg for decades, when he observed the dire impact of timber shortages on the metallurgy industries. For him, "sustainable use" of a forest can only be achieved if one refrained from extracting more wood than can be regrown through reforestation management and without destroying the precious resource in the long run. Current public debates on energy transition and climate change have claimed a straightforward causal link to Carlowitz's work and his term *Nachhaltigkeit*—coined at a time when enlightenment was still in its infancy and mercantilism rather than modern capitalism ruled economic affairs. Yet, there has never been a direct link from Freiberg in 1713 to Rio in 2012, from *Sylvicultura oeconomica* to the recent *Report to the Club of Rome 2052*.[7]

Moreover, as historian Richard Hölzl explains, the emergence of sustainability in German scientific forestry has been a contested story from the start: "Focusing on timber production and financial revenue for the state treasury, scientific forestry simplified the biological composition of forests, re-organized their internal structure along the lines of legibility and accountability, and restricted access for users other than scientifically trained personnel."[8] For one, the scientific mode of forest management met local resistance and clashed with the vested interests of other groups in society. It turned out that "sustainable" forest management increased the vulnerability of forest environments to droughts, storms, and forest pests. In the tradition of Carlowitz, sustainability transformed nature into a commodity that could be measured, registered, accounted, and taxed.

Sustainability in the sense of turning nature into a commodity promoted ideas of rationalizing and standardizing the natural world, as James C. Scott observed in his seminal study on the emergence of modern statecraft.[9] Furthermore, such a high-modernist viewpoint that sees the world through the eyes of state power clashes with the widely accepted definition of sustainability as "meeting the needs of the present without compromising the ability of future generations to meet their own needs"—a formulation we owe to the United Nations' 1987 Brundtland Commission. Neither Gro Harlem Brundtland nor Dennis L. Meadows and his coauthors of the *Limits to Growth* report for the Club of Rome of 1972 knew anything about Carlowitz. In forestry science, Carlowitz's concept of a sustained yield continued to be highly esteemed internationally: the 1951 UN Food and Agriculture Organization (FAO) report, entitled *Principles of Forest Policy*, states that it would take another two decades before the internationally accepted term of the forestry profession "was to serve as the blueprint for the universal concept of 'sustainable development,'" as historian Ulrich Grober pointed out.[10]

The claim for a straightforward history of the term *ecology* is equally problematic. As Robert Friedel's essay reminds us in this volume, the term experienced a similarly long, nonlinear history of creation and transformation. When, in 1866, German naturalist Ernst Haeckel coined the term *ecology* (Ökologie), he linked the maintaining of order in human households and communities to that of the Earth's environment: Planet Earth needs care like a home does. It took more than a century before ecology—fully stripped of its post-Darwinism roots—could develop into a more rigorous scholarship that links evolutionary biology and environmental sciences together to analyze the interaction between living things and their environment.

The concept of ecology was first embraced by UNESCO's Man and the Biosphere Program in 1970. Two years later, the UN Conference on the Human Environment in Stockholm established a set of principles aimed at strengthening Earth's capacity to produce renewable resources. The late 1960s and early

1970s saw not only intergovernmental top-down ambitions for safeguarding the planet, but also a rich palette of bottom-up movements, many of which—including Steward Brand's Whole Earth network and Greenpeace, founded in 1972 by a group of Californian hippies—emerged from a remarkable fusion of countercultural movements and technoscientific expert communities.[11] The rise of these new environmental movements marked an important turning point in environmental history. The movements were also a response to the transition from a slow-moving to a rapid loss of global sustainability that had begun already in the 1950s.[12]

In short, despite efforts to establish lineages to earlier times, the 1970s still seem important as a turning point.

Toward a Nonlinear, Cyclical History of Sustainability

How should we interpret a movement that explicitly sought to resurrect older practices for new environmental purposes? The issue of what constitutes a turning point in history—indeed how change occurs—has been subject to debate. Two scholarships are of importance in our discussion of how we need to understand the story of cycling and recycling in a larger historical timeframe. One has resulted from innovation studies, the other from environmental history.

Recent innovation studies have come to appreciate so-called enduring technologies—those used daily and almost casually discovered rather than the capital-intensive ones invented in research and development labs. It helps us to understand the key actors of the 1970s—rebelling consumers—who viewed cycling and recycling as acts of green citizenship. They revived cycling as a mode of sustainable transportation and advanced these relatively low-tech and low-carbon technologies as innovative tools for sustainable mobility and resource management. Their impact has been profound: today, many urban policy makers have come to embrace bicycles as their favorite mobility policy instrument for more livable and sustainable cities. In the same manner, policy makers have focused on waste recycling as a cornerstone in dealing with the planet's limited resources. Given the enormous negative associations of these technologies as old-fashioned and antimodern, the grassroots and policy success has been a remarkable turn of events. More importantly, these practices challenge easy narratives of innovation as a series of progressive steps.

The strand of innovation scholarship argues that stories of use, rather than invention and innovation, demonstrate the enormous significance of these relatively low-tech technologies in people's daily lives; therefore, they also should be central to understanding innovations. Historians of technology such as David Edgerton first issued a call to decenter innovations as the premier site of technological progress.¹³ This insight has now also reached innovation studies, which theorize how policy makers can best introduce environmentally beneficial innovations, in situations where stakeholders have a vested interest in keeping old and unsustainable systems intact. Traditionally, these theorists have concentrated on transitions and tipping points, exploring how entrenched systems like our dependence on unsustainable fossil-fuel economies can move to more sustainable economies most effectively.¹⁴ Given that change is a complex issue, these theorists of sustainability have sought to learn from historical scholarship to advance their own inquiries.

The British sociologist of technology Elisabeth Shove, in particular, has turned to the historical scholarship of cycling to explain why examining older technologies is theoretically important for environmental studies. Innovation studies and transition theory successfully explain when and how innovations have come into existence and gelled into systems, but they pay less attention to how old innovations were maintained or revitalized because, she suggests, focusing on older technologies is detrimental to the dominant narrative of progress. Scholars tend to concentrate on "processes of emergence and stabilization" rather than on "those of disappearance, partial continuity, and resurrection." When analyzing innovations, we should focus instead on understanding how they have been shaped by persistence, continuity, and the revival of old technologies: "How dormant remains of past regimes come back to life and how innovation journeys start over again." Using the historical case study of cycling to make her theoretical point, Shove suggests to "set the terminology of replacement and substitution aside and concentrate instead on how cycling and driving are positioned [in relation to one another], as their trajectories develop and decline." She concludes that the successful resurrection of old technologies is based on "pockets of persistence," rooted in (still) existing materiality, know-how (expertise), user routines, and an active new cultural framing that fits new contexts.¹⁵

Indeed, historians—specialists in examining the dynamics of change—are particularly well equipped to focus on such pockets of persistence. In the analysis of how developments come about, historians have a useful toolbox at their disposal to examine issues of continuity and discontinuity, of developments that endure and those that have been ruptured. In this volume, Georg Stöger refers to the long tradition of secondhand trading, dating way back to the early modern period, which he interprets as practices of recycling. Roman Köster stresses the ruptures in organizational structures and technological cultures of recycling in West Germany after the Second World War. Technology users are often the carriers of pockets of persistence, as Djahane Salehabadi points out in her case study on the battle over the waste stream and urban mining, again in West Germany. Users are also in the business of launching protests and resistance to system builders who lobby for new systems. Technology users have played an important role in the survival and reappearance of the "old" cycling and recycling technology in the environmental movement. Indeed, in the West, such pockets of persistence turned into movements of resistance in the 1970s.

Political scientists suggest that movements need social organizations to achieve their well-defined goals. In terms of power relations, political opportunities also need to be conducive for activism to blossom into movements; it includes greater access to political decision-making power and to elite allies as well as the growing instability of ruling elites and the state's declining capacity to repress dissent. In cultural terms, political scientists now recognize that the act of framing an issue is important in helping activists to mobilize potential recruits and audiences like the media, elites, and sympathetic allies.[16] What we have learned from this scholarship is that, by the same token, in order for older technologies to become viable again, they need a movement's social organization, political leverage, and cultural framing. What made the 1970s particularly successful and different was the combination of these three important elements: its broad-based social movement, its transnational political coalition building, and its fundamental cultural reframing. These insights may help us understand the seemingly simple question of why older technologies such as bicycles and recycling became popular and legitimate again during the 1970s. During this era, everywhere in the Western world, environmental activists began to recycle as a political act in a broad-based social movement.[17] Similarly, bicycle activists in the 1970s sought to build a large social movement—a critical mass—to change mobility policies.[18]

In the cases of both cycling and recycling, the cultural (re)framing in the pivotal decade of the 1970s proved essential in making change possible at all. In both cases, this cultural reframing was quite a tour de force. For decades, bicycles had been negatively associated with working class rebellion, chaotic cities, and undisciplined behavior.[19] War also generated a negative discursive place for bicycles, as Catherine Bertho Lavenir shows in her contribution. Yet, in their roles as environmental activists, urban-based consumers came to reframe the bicycle as the ideal vehicle to meet the new social challenges for sustainable, silent, clean, safe, cheap, and efficient urban transport. Only by bringing bicycles into the discourse of modernity and speed was the Human-Powered Vehicle movement able to recast bicycles as a site of innovation, as Manuel Stoffers explains in his chapter. By changing the image of bicycles from a working-class vehicle to a desirable tool for green citizenship, cycling gained a fighting chance for equal treatment among motorized traffic, when funds for infrastructures and urban planning were allocated, as Ruth Oldenziel and Adri Albert de la Bruhèze argue in their essay on the history of bicycle taxes.

At the same time, it also became increasingly evident that while social organizations like the environmental moments are crucial for social change, technological systems endure and create path dependencies that are hard to break—they become the true "monuments of unsustainability," as Martin Emanuel makes clear in his essay on urban planning and cycling in Stockholm.[20] Social organization, cultural reframing, and political leverage may not be enough for old technologies to be successful again. The practices of recycling and cycling are up against large entrenched technical systems that carry weight and momentum that are hard to change. Indeed, we need to recognize that cultural reframing and political leverage may not be enough to undo the kind of path dependency that is institutionalized in large technical systems and have become the monuments of unsustainability. These contributions caution us to understand the story of innovation as a simple linear process of progress.

Well-intentioned policy may have unwanted and even disastrous outcomes. As Zsuzsa Gille reminds us, in East Europe the political transformation from socialist to postsocialist societies within the European Union policy framework established conditions leading to environmental disasters rather than preventing them; and as Bill Steele points out, creative attempts in Japan to recycle abandoned bicycles have ended up aggravating environmental problems rather than solving them. Even more importantly, Hans Peter Hahn's contribution on bicycles in Africa reminds us that Western narratives of change are limited in tracing a sustainability discourse and practice outside the Western beliefs in material progress on the one hand, and the industrial development as an inevitable march forward into resource exhaustion on the other.

The second strand of scholarship that has questioned the unilinear progression of change comes from environmental history. Recent scholarship in environmental history has mobilized a far larger timeframe that goes beyond the discussion of whether we need to see the 1970s as a pivotal turning point along the path of historical time, or whether the Western narrative is limited in capturing the stories from Africa. Environmental history has, moreover, embarked on questioning the linearity of change in nature-culture-relations that often have been told as stories of decline, decay, and degradation.

Today, most observers agree that humanity has become a global factor that affects the overall Earth system in sectors such as water circulation, climate, biodiversity, sedimentation patterns, and use of lands and seas. To pay tribute to the deep impact we as humans have on the environment, a conceptual framework has been proposed that would transcend the sustainability paradigm: the term *Anthropocene*, which was popularized by biologist Eugene F. Stoermer and Nobel Prize–winning atmospheric chemist Paul J. Crutzen around the year 2000. The core thesis is that humanity has affected nature in such a way that a new, human-made stratum has emerged in the geological record. Only a few years after Crutzen and Stoermer popularized the Anthropocene as the new geological "age of mankind," the International Subcommission on Quaternary Stratigraphy established a working group to determine whether there is enough scientific evidence to define a new Earth era. This new era, the Anthropocene, would succeed the Holocene. The Greek word *holocene* literally means "entirely recent," which indicates there is not much room for moving to something novel in a discipline that usually counts in hundreds of thousands and millions of years.

While earth scientists discuss the hard facts of geological strata, historians have started a lively debate about periodizing the Anthropocene.[21] Three periodization schemes have been proposed. The first is the Neolithic Revolution, which began about 11,000 years ago when humans started to use agriculture in addition to hunting and gathering. Second is the Industrial Revolution that started in Great Britain in the late eighteenth century. The final periodization scheme is the Great Acceleration at the beginning of the second half of the twentieth century, when almost all parameters of human intervention in nature changed from linear to exponential growth. In all three schemes, technology plays a prominent role. Novel technical solutions spurred the transition from societies based on hunting and gathering to agriculture and settlement; mechanization and the transition from renewable energy resources to fossil fuels spurred industrialization; and the Great Acceleration was driven by consumers' mass use of technologies.

Environmental history not only has broadened issues of periodization beyond narrations of industrial development and economic progress. It has also fundamentally questioned the unilinear notions of history to reassess older notions of cyclical interpretations of history. In that context, we have come to appreciate that the very notion of the future as an undetermined space that is open to human creativity is a recent invention. Only around the long transition to the nineteenth century, when the enlightenment finally gained ground in Western societies, were cyclical ideas of futures that were bound to Christian eschatology dismissed. In its stead, the "discovery of the future"—the singular is crucial—became an integral part of the Western project of modernity. At the time, the belief in the future helped transform history from a cyclical into a linear endeavor.[22] Henceforth, in the professional domain and the public realm, "history" came to be seen as a linear mode of succession of change. Periodization became the noble and central task of the historians' profession; graphical tools such as timelines and chronologies fostered a linear understanding of history.[23]

The Western idea of future as a linear project has become increasingly contested—a trend that has been reinforced by the success of postmodernism, postcolonialism, and globalization scholarship. The field of technology, where the idea of endless progress and a linear concept of innovation was particularly deeply embedded, has at last been affected by these trends as well, as indicated by the scholarship of Shove and others. Historians and sociologist of technology have struggled against linearity and the hegemony of modern forms of one-dimensional futures. From such a perspective, history of technology has been understood as an open source of knowledge that provides orientation in current debates about the present and the future by uncovering creative ideas buried in the past.

In its appreciation of cyclical forms of historic progression over linear models, this volume is taking the idea of the openness of both the past and the future seriously. In doing so, the individual chapters emphasize the fact that *recycling* often means repurposing and reimagining. This also holds true for cycling. The widely debated concepts of "cradle to cradle" and "upcycling," which Michael Braungart and William McDonough have developed to rethink recycling as a sustainable mode of reusing things and stuff, may still fall short in stressing the cyclical dimensions of material flows in societies.[24] Yet they both point to the potential of a nonlinear understanding of sustainability—and they stress the need to pay tribute to economic factors. As Donald Worster reminds us in his essay, the quest for sustainable technologies will fail altogether if it neglects to question its very foundation and belief in economic growth as the underlying model for sustainability, no matter how many bottles we recycle or how many bicycles we ride. In this debate, the insights from historical scholarship on cycling and recycling may serve to better contextualize our current debates on the transition to a more sustainable society.[25]

Notes

1. Ruth Oldenziel and Mikael Hård, *Consumers, Users, Rebels: The People Who Shaped Europe* (London, 2013), 235–271. We thank Adri A. de la Bruhèze, Mikael Hård, and Gijs Mom for their suggestions and comments. Special thanks goes to Katie Ritson, Marielle Dado, and Brenda Black for their invaluable editorial support.
2. Fred Turner, *From Counterculture to Cyberculture: Stewart Brand, the Whole Earth Network, and the Rise of Digital Utopianism* (Chicago, 2006).
3. Portola Institute, "Purpose," in *Whole Earth Catalog: Access to Tools* (1969).
4. Turner, *From Counterculture to Cyberculture*.
5. Garrett De Bell, "Recycling," in *Environmental Handbook*, ed. J. Barr (London, 1970), 217, as quoted in Timothy Cooper, "War on Waste: The Politics of Waste and Recycling in Post-war Britain, 1950–1975," *Capitalism, Nature, Socialism* 20, no. 4 (2009): 53–72, here 62. *Consumers' Guide* was published in 1971. Andrew Jamison, *The Making of Green Knowledge: Environmental Politics and Cultural Transformation* (Cambridge, 2001), chapter 4.
6. Carl von Carlowitz, *Sylvicultura oeconomica oder Haußwirthliche Nachricht und Naturmäßige Anweisung zur Wilden Baum-Zucht*, ed. Joachim Hamberger (Munich, 2013 [1713]).
7. Ulrich Grober, "Von Freiberg nach Rio–Carlowitz und die Bildung des Begriffs Nachhaltigkeitin," in *Die Erfindung der Nachhaltigkeit. Leben, Werk und Wirkung des Hans Carl von Carlowitz*, ed. Sächsische Carlowitz-Gesellschaft (Munich 2013), 13–30; and Franz Josef Radermacher, "Die Ressourcen der Erde setzen uns Grenzen—vom säch-

sischen Bergmann Hans Carl von Carlowitz 1713 bis zum neuen Report an den Club of Rome 2052," in ibid., 141–156.
8. Richard Hölzl, "Historicizing Sustainability: German Scientific Forestry in the Eighteenth and Nineteenth Centuries," *Science as Culture* 19, no. 4 (2010): 431–460, here 431.
9. James C. Scott, *Seeing Like a State: How Certain Schemes to Improve the Human Condition Have Failed* (New Haven, CT, 1998), 2–4.
10. Grober, "Von Freiberg"; see also Ulrich Grober, *Sustainability: A Cultural History* (Totnes, 2012).
11. Frank Zelko, *"Make it a Green Peace": The Rise of Countercultural Environmentalism* (New York, 2013).
12. Christian Pfister, "The '1950s Syndrome' and the Transition from a Slow-Going to a Rapid Loss of Global Sustainability," in *The Turning Points in Environmental History*, ed. Frank Uekötter (Pittsburgh, PA, 2010), 90–118.
13. David Edgerton, *The Shock of the Old: Technology and Global History since 1900* (London, 2006). Oldenziel and Hård, *Consumers, Users, Rebels*. See also Nelly Oudshoorn and Trevor J. Pinch, eds., *How Users Matter: The Co-Construction of Users and Technology* (Cambridge, MA, 2003).
14. Frank Geels and Johan W. Schot, "Typology of Sociotechnical Transition Pathways," *Research Policy* 36 (2007): 399–417.
15. Elizabeth Shove, "The Shadowy Side of Innovation: Unmaking and Sustainability," *Technology Analysis & Strategic Management: Innovation, Consumption, and Environmental Sustainability* 24, no. 4 (2012): 363, 367, 372–373.
16. On the social movement debate, see for example Marco Giugni, Doug McAdam, and Charles Tilly, *How Social Movements Matter: Social Movements, Protest, and Contention* (Minneapolis, MN, 1999); Jeff Goodwin and James M. Jasper, *Rethinking Social Movements: Structure, Meaning, and Emotion, People, Passions, and Power* (Oxford, 2004).
17. On environmentalism as a broad-based social movement, see for example Adam Rome, *The Genius of Earth Day: How a 1970 Teach-In Unexpectedly Made the First Green Generation* (New York, 2013). Rome, *The Bulldozer in the Countryside: Suburban Sprawl and the Rise of American Environmentalism* (Cambridge, 2001). For transnational coalition building, see Jan-Henrik Meyer, "Saving Migrants: A Transnational Network Supporting Supranational Bird Protection Policy," in *Transnational Networks in Regional Integration: Governing Europe, 1945–83*, ed. Wolfram Kaiser, Brigitte Leucht, and Michael Gehler (London, 2010), 176–198. For practices of recycling, see Ruth Oldenziel and Milena Veenis, "The Glass Recycling Container in the Netherlands: Symbol in Times of Shortages and Abundance, 1939–1978," *Journal of Contemporary European History* 22, no. 3 (2013): 453–476; Andrea Westermann, "When Consumer Citizens Spoke Up: West Germany's Early Dealings with Plastic Waste," *Journal of Contemporary European History* 22, no. 3 (2013): 477–498.
18. Zachary Mooradian Furness, *One Less Car: Bicycling and the Politics of Automobility* (Philadelphia, PA, 2010).
19. Adri A. de la Bruhèze and Frank C. A. Veraart, *Fietsverkeer in praktijk en beleid in de twintigste eeuw. Overeenkomsten en verschillen in fietsgebruik in Amsterdam, Eindhoven, Enschede, Zuidoost Limburg, Antwerpen, Manchester, Kopenhagen, Hannover en Basel* (Eindhoven, 1999).

20. See Martin Emanuel in this volume; Langdon Winner, *The Whale and the Reactor: A Search for Limits in an Age of High Technology* (Chicago, IL, 1986); Ruth Oldenziel and Adri A. de la Bruhèze, "Contested Spaces: Bicycle Lanes in Urban Europe, 1900-1995," *Transfers* 1, no. 2 (2011): 31–49.
21. Will Steffen, Paul J. Crutzen, and John R. McNeill, "The Anthropocene: Are Humans Now Overwhelming the Great Forces of Nature?" *Ambio* 36 (2007): 614–621; Will Steffen et al., "The Anthropocene: Conceptual and Historical Perspectives," *Philosophical Transactions of the Royal Society Academy A* 369 (2011): 842–867; John R. McNeill, Will Steffen, and Paul Crutzen, "The Three Stages of the Anthropocene," *Island GeoScience* 5, no. 1 (2008): 2–5; Dipesh Chakrabarty, "The Climate of History: Four Theses," *Critical Inquiry* 35 (2009): 197–222; Helmuth Trischler, ed., *Anthropocene: Exploring the Future of the Age of Humans* (Munich, 2013); Nina Möllers, Christian Schwägerl, and Helmuth Trischler, eds., *Welcome to the Anthropocene: The Earth in Our Hands* (Munich, 2015).
22. Reinhard Koselleck, *Futures Past: On the Semantics of Historical Time* (New York, 2004), and Lucian Hölscher, *Die Entdeckung der Zukunft* (Frankfurt, 1999); see also Niklas Olsen, *History in the Plural: An Introduction to the Work of Reinhart Koselleck* (New York, 2012). Peter Burke has emphasized that the late middle ages and particularly the seventeenth century saw multiple "future-oriented practices," but he also has admitted that the idea of a more distant future as a space open to human creation is a project of modernity; Peter Burke, "Foreword: The History of the Future, 1350–2000," in *The Uses of the Future in Early Modern Europe*, ed. Andrea Brady and Emily Butterworth (New York, 2010), ix.
23. Daniel Rosenberg and Susan Harding, eds., *Histories of the Future* (Durham, 2005); Daniel Rosenberg and Anthony Grafton, *Cartographies of Time: A History of the Timeline* (Princeton, NJ, 2010).
24. William McDonough and Michael Braungart, *Cradle to Cradle: Remaking the Way We Make Things* (New York, 2002), and McDonough and Braungart, *The Upcycle* (New York, 2013).
25. See also Christof Mauch and Helmuth Trischler, *International Environmental History: Nature as a Cultural Challenge* (Munich, 2010); Bernd Hermann and Christof Mauch, eds., *From Exploitation to Sustainability? Global Perspectives on the History and Future of Resource Depletion* (Stuttgart, 2013).

 PART I

Cycling Histories

 CHAPTER 1

Use and Cycling in West Africa

Hans Peter Hahn

"The *Man with a Bicycle* is produced by someone who does not care that the bicycle is the white man's invention—it is not there to be Other or the Yoruba Self; it is there because someone cared for its solidity; it is there because it will take us further than our feet will take us; it is there because machines are now as African as novelists—and as fabricated as the kingdom of Nakem."— Kwame Anthony Appiah, 1992[1]

In many areas in Africa, bicycles are a prominent part of the busy streets of urban agglomerations as well as the narrow paths of rural areas.[2] Bicycles are highly esteemed, sometimes even prestigious consumer goods; they are also objects of everyday necessity and in some contexts even publicly despised. This double relevance applies to the retail traders in town, who use bicycles to transport their goods from the periphery to the central market, to pupils in rural areas, who would not be able to make it to the nearest school without them, and to the peasants, who use them to reach their bush fields dozens of kilometers from their homesteads. As in many places of the world, bicycles in Africa are important for the generation of income, owing to the demand for the transport of loads and people (i.e., *bode bode,* or taxi-bikes).[3] Bicycles are not taken for granted: for many, particularly in rural areas, they are objects of dreams, as economic constraints make the acquisition of one difficult. Due to limited budgets, the purchase of a bicycle constitutes a major investment. Furthermore, difficult road conditions and the need for regular maintenance require the frequent mobilization of resources for long-term use of bicycles.

This chapter focuses on the appropriation of bicycles in local societies in West Africa. It foregrounds local usages and meanings, and, related to this, addresses both matters of transportation and the status of the bicycle as a luxury item.[4] Although this contribution is not a historical overview, it will start with some historical sketches. The bulk of data presented in the following section is ethnographic information collected by the author in Burkina Faso and neighboring countries. More broadly, bicycles exemplify the material and contextual modifications that global goods undergo when they first enter into any cultural space. This process is known as "cultural appropriation," and I use it to describe the creation of new uses for the bicycle, its interpretation in local contexts, and its situatedness with regard to other tools and activities.[5] As expressed in this essay's epigraph by the philosopher Kwame Anthony Appiah, bicycles are an outstanding example of the impact of so-called modern and Western technologies in Africa. This chapter will therefore approach bicycle use and its meanings in terms of a double transformation: in which bicycles change people and societies, and in which bicycles, like many other global commodities, are modified by their contextualization.[6]

The conceptual framework of cultural appropriation helps us to consider bicycles as part of a transformative process mastered by bicycle users and technicians in West Africa. In deliberately adopting a perspective that puts people's activities in the foreground, this chapter intends to highlight creativity at the grassroots level. Cultural appropriation adopts a perspective on transformations authored by the bicycle users themselves, and not so much the various projects' action plans and implementation schemes. The concept of cultural appropriation contributes to overcoming the questionable dualism of the "Africanized" bicycle's "African" or "modern" character. Like many other ubiquitous everyday technologies, such as mobile phones and plastic sandals, bicycles in Africa are part of a specific "African modernity."[7]

In the light of the ubiquity of bicycles in many areas in Africa, it is astonishing that, until recently, scholarly research on African societies has paid little attention to this phenomenon and its related activities.[8] I suggest two specific reasons for this neglect. First, for historians and social anthropologists who are interested in local traditions, the bicycle does not seem to be something typical of Africa; it does not appear to constitute a fundamental part of any "African culture." Second, for social scientists and experts in transportation conducting their research for a better future of African societies, the bicycle has only lately been seen as a potential contributor to development. Grassroots development activities pointed as early as 1976 to the relevance of bicycles, whereas large-scale projects promoting bicycles appeared only in the past two decades.[9]

Peter Cox, in *Moving People: Sustainable Transport Development,* presents a comprehensive overview on numerous development projects regarding non-motorized transport (NMT) in Africa.[10] My approach is much in line with Cox's arguments, and differs from developmentalist perspectives on transport issues in Africa by referring mainly to ethnographic data, by putting the agendas and reports of these projects in the foreground, and by focusing on the agency of African bicycle users. Bicycle adaptation may be related to economic issues, as frequently highlighted in development reports, but it is considered here in the first place as a cultural phenomenon, including tradition, history, and innovation through appropriation. This perspective elucidates the embedding of bicycles as a particular form of mobility in society, but also as a contested topic of prestige. The recent celebration of the bicycle as an element of the intermediate means of transport (IMT) by advocates of development agencies (in particular by the World Bank) has provided ample economic evidence for the usefulness of the bicycle, given the limited means of farmers in Africa. The bad road conditions common in African rural areas anchor a further argument in favor of using bicycles.[11] However, this study does not intend to dwell on the economic advantages of bicycle use, but rather intends to highlight the bicycle's embeddedness in everyday culture.

The historical scenes of bicycle usage in Africa described in the next section will highlight the specificity of this means of mobility. The larger section that follows will provide more specific information about the current usage of bicycles in Burkina Faso, where the author conducted extended ethnographic research.

Early Documents on Bicycles in West Africa: The Starting Point of Cultural Appropriation

The history of the bicycle in Africa is an almost entirely neglected field in the domain of the history of technology as well as in the colonial history.[12] This gap cannot be filled within the limited space of this chapter. However, it is useful to sketch the early years of bicycles in Africa.

Bicycles arrived in Africa more or less simultaneously with their introduction as a consumer good in Europe. It is not misleading to consider the earliest historical period (before 1914), when the bicycle in Africa was as much "up-to-date" as in Europe, as the starting point of its appropriation. We can consider on a more general level that the history of bicycles in Africa is a history of innovation. But, very much in line with what has been said above, it is not so much about technological innovation, but rather innovation on the level of adaptation to the customers' and users' requirements. More precisely, *adaptation* might be the wrong term, because, as shall be shown in the following, there has never been a supplier modifying the models according to what is required, but rather the users do this themselves. The action taken by the bicycle users and the intentional enhancements they assigned to the available bicycles are conceptually framed by the term *cultural appropriation*.

Thus it is legitimate to say that the process of appropriation started with the availability of bicycles in Africa that were physically identical to what was in use in Europe at that time. If today some experts argue for the production of specific bicycles for Africa, this should be considered as a consequence of specific practices and patterns of usage of bicycles there; it is an outcome of cultural appropriation.

A further condition of cultural appropriation is the interest and valuation of the item to be appropriated. Three short references may contribute to give some evidence for this. They originate from the first decade of the twentieth century, and thereby constitute some of the earliest records of bicycle use in West Africa. These references are closely connected to the colonial times, and are most likely biased by the related ideology, though discussion of these biases is beyond the scope of this essay.

The first report is from the missionary physician Rudolf Fisch, who in the years 1909–1910 traveled more than 2000 kilometers through the British colonies of Gold Coast and Togo. Fisch considered long-distance traveling by bicycle in Africa a new form of mobility, one which allowed him to point polemically to the detrimental effects of the expeditions dominantly practiced at that time.[13] As he wrote, his new means of transport, which did not require carriers, enabled him to come into direct contact with the villagers and expand the options for communication. Fisch furthermore highlighted the fact that he and his African companion both made the whole trip on their bicycles without any technical problems. Wherever they arrived, the bicycle gave rise to astonishment and curiosity: "Wherever we appeared with our bicycles ... many curious faces reappeared and observed from their secured hiding place how these strange people (who we were in their eyes) acted with these uncannily shining machines.... A number of younger men testified their joy about the presence of our bicycles by running the way along ahead of us."[14]

The second report is from Hans Schomburgk. During the years 1911–1912, he traveled in West Africa as a private entrepreneur with the objective of catching rare mammals for European zoos. He chose the bicycle for its ease of mobility and communication. With his bicycle, he could make spontaneous detours and gather information on animals from villagers. The perception of the bicycle by the local population is described as follows: "[After the carriers had left] I was the last who stayed in the village. Several people came in order to say good bye and to admire the white miraculous animal with the bicycle. For them this machine was a fetish, a new wondrous thing, about which even the most experienced men ... could not tell anything, because over there were no bicycles at that time."[15]

The two travelers share an important feature: both condone being associated with the colonial authorities, while still following their own agendas and objectives.[16] By no means do they promote resistance to colonial power, and they willingly accepted the support of the colonial administration if it were of use to them.

An image taken in Lomé in 1911 by an African photographer, Alex Acolatse, complements the perspective on bicycle uses in Africa during the early colonial phase. Acolatse was among the well-known personalities in town, having a shop in the city and selling pictures to Europeans and to African clients. Acolatse's photograph included the following caption: "Bicycle race of the indigenous population at the anniversary of the Emperor in 1911." The image indicates that cycling was an acknowledged sport for Africans as well as Europeans. Whereas in the two first reports, the bicycle users are white men who fascinate the Africans with their machines, the third spotlights Africans themselves as actors, ready to become masters of the bicycle.[17]

Against the background of these stories and images, the popularity of bicycles in Africa becomes clear.[18] A report from Nigeria further documents how bicycles achieved the status of an important means of transport for merchandise during the 1930s. Competing with the entrepreneurs' lorries, young men used their bicycles to transport the locally produced palm oil to the harbor towns. The cyclists were not only successful in creating a beneficial business with their bikes: they also founded "bicycle clubs," repair shops, and credit circles. In 1934, an "army" of more than 20,000 cyclists, carrying oil drums, threatened the established motorized transport. The colonial administration, anxious about the autonomous organization of the cyclists, made plans for a special bicycle tax to reduce the profits of the nonmotorized transports.[19] However, this was not realized, because the administration recognized that the cycling transport business was particularly valuable during the Second World War, when fuel shortages became more and more frequent.

As these reports make clear, specific functions had been assigned to bicycles in Nigeria, and Africa more generally, at that time, and bikes had been locally modified in order to serve these purposes. The bicycles were imported from England and assembled in Nigeria. Local mechanics added sturdier baggage carriers, appropriate for the transport of the heavy oil drums. The standard model at that time was the Raleigh Roadster, with a black diamond frame and rod brakes. This model had been used in England during the First World War, but it was assembled from the 1930s onward also in India and China.[20] Raleigh also developed a specific advertising campaign, showing an African riding his bicycle to escape a lion.[21] As documented in Gregory H. Bowden's history of the company, the United Africa Company (UAC) was an authorized Raleigh distributor in West Africa.[22] As early as 1946, Nigerian traders were present in Shanghai, seeking to make contracts with Chinese manufacturers for the import of bicycles or bicycle parts from China to West Africa.[23] Perhaps these initiatives were undertaken in order to undermine the monopoly of the UAC. Other sources make clear that the UAC had a manifest interest in keeping a monopoly on the importation of bicycles in several African countries.[24]

In the year 2000, the Chinese Phoenix company exported more than a million bikes per year in more than fifty countries worldwide. For Africa, Phoenix produces a slightly modified version of the Roadster.

Bicycles in Africa

Although the bicycles used in Africa are not locally manufactured, there are good reasons to call them "African bicycles." The following section will justify this claim by describing the modifications as an outcome of cultural appropriation. This evaluation is based on ethnographic observations made in 2001–2003 in Kollo, a village in southern Burkina Faso, in the department of Tiébélé. In order to make the transformation of the bicycle as evident as possible, the starting point will be the sketch of an "ideal biography" of a bicycle.[25]

If someone in this rural community has managed to amass the necessary amount of money to buy a bicycle, he will walk the distance of 30 or 40 kilometers to one of the larger markets in the region. There he can find several bicycle dealers who exhibit imported bicycles in a special area within the marketplace. In order to make sure that the bicycle is really new, the traders leave some of the paper wrap on the frame. After purchase, the new owner will not remove these fragments of packaging, as they display to the public the mint condition of his possession. Further obligatory attributes of new bicycles are a pump, saddle cover, and lock—the bicycle owner acquires the last two items separately. The lowest price for a new bike is less than the equivalent of €100, which at first glance might appear to be a modest sum. However, this is more than the annual budget of many rural households (no more than €70). People are able to live on such little income only because staple foods (millet, maize, etc.) are produced by members of the household for their own consumption.

Against this background, it is understandable that new bicycles are treated as precious objects. During the first weeks after acquisition, they are not lent out and their use is limited to visits to the market. This is a highly meaningful activity, as markets in Burkina Faso are key places for building an individual's social reputation; they are also forums for the exchange of news and gossip, which includes information about who owns a new bicycle and how they managed to acquire such an expensive item. Very often, the ownership of a bicycle is related to a history of labor migration, and there are reports of how the migrant workers returned with pockets full of money or with loads of consumer goods. In many cases, the migrant hands over his recently acquired bicycle to his brother or uncle to leave a sign of his presence in the village, although he himself will not stay more than a few weeks or a month. There are only a few cases where a bicycle had been acquired with the limited means of one's own harvest of maize, rice, or vegetables. As people in the village say, "working on the fields does not pay for a bicycle."

A few weeks after its acquisition, the bicycle is no longer merely an object of wealth and status; rather from that time onward, its primary value lies in its uses. Among the most important and frequent destinations of everyday trips are the distant bush fields, markets, and, for pupils, schools. The bicycle is also instrumental in bringing sick persons to the nearest hospital, among other miscellaneous transports. It is said that for a proud owner of a bicycle, who uses it for all purposes, "his bike has become his feet." The high intensity of usage is underlined by the fact that a socially respected bicycle owner will not hesitate to lend his bicycle to his neighbors and relatives when he does not need it himself. Whoever needs it (whether because they don't own one, or because theirs is under repair) may ask for it. It is of course possible to reject such a request, but this would be socially problematic for the bike owner. Furthermore, the owner would run the risk of not being able to ask for a neighbor's bicycle if they were in need.

Usage necessitates considerable changes to the material appearance and functionality of the bicycle. Soon after the initial phase, the owner starts to remove some parts and add others. Some bicycle parts are almost immediately removed because they are considered useless or even dangerous: the chain guard, the mudguard, the rear light, and the front brake. People are well aware that these parts are visible signs of a "new bicycle," but, simultaneously, they consider them superfluous to its functioning. Often the intention is to remove these parts only temporarily in the process of changing a tube or the chain, but during the repair, the owner realizes that it is complicated to reassemble the parts, or that the screws required for doing so have been lost or used otherwise. It seems easier then to use the bicycle as is. From the perspective of the user, the bike becomes even more secure: loose or incorrectly mounted parts might block or even damage a wheel during a trip. Frequent repairs (like changing the tire and tube or mounting the chain) become easier when the "useless" parts are removed for good. It is easier to remove a wheel when there are no fenders, and the chain is much easier removed and replaced if there is no chain guard. The rear light is just a potential cause of electric short circuits. Last but not least, the front brake is considered as a source of potential accidents on gravelly ground; once it is removed, it is safer not to put it in place again.

Jojada Verrips and Birgit Meyer have dealt with these kinds of modifications and repairs in the context of motorcars in Ghana. They call the process "tropicalization," and consider these intentional changes the outcome of a particular competence with regard to the technology of the vehicle and of the ability to realize robust solutions with limited means.[26] This perspective is similar to Kurt Beck's observation regarding diesel engines in the Republic of Sudan. These motors are used to pump water from the river Nile to the irrigated fields nearby. Beck calls this a "technological culture of poverty," which affirms that there is a culturally integrated knowledge about technology without formal education, like engineering, and without standardization and measuring instruments.[27] Thus the cooling of the motor must be guaranteed even if there is neither a temperature sensor nor an indicator of the motor temperature in the car. As Verrips and Meyer report, the driver can recognize by the sound and the smell of the exhaust fumes whether the motor is overheating or not. In the case of the bicycle in Burkina Faso, the driver can recognize from the shaking of the bicycle and its noise whether a rim is deformed or the chain has loosened.

These skillful ways of monitoring the healthy state of a bicycle—as with a car or water pump—reflect pragmatism and improvisation among bicycle users. Many changes are simply the result of lacking spare parts, screws, or tools. The current state of the machine is considered as something temporary that is intended to serve for a limited time until the next revision. With the passage of time and ongoing use, many improvisations prove to be durable and reliable and the owner comes to terms with the state of his vehicle.

The provision of spare parts is a permanent challenge in many regions in Africa, even if there are traders that specialize in bicycle parts with almost complete assortments, including tubes, tires, and rims of different sizes, as well as spokes, cranks, pedals, and even ball bearings. In many situations, however, the owner of the bicycle lacks the financial means to buy the necessary parts, which can lead to the temporary demobilization of the bicycle. A frequent strategy used to avoid this scenario is to borrow needed parts from neighbors whose bicycles might also be temporarily unusable. A typical bicycle biography might go as follows: after a phase of active use, the bicycle is waylaid by a minor defect. Then the lack of means or the generosity of the owner leads to the further dismounting and lending out of parts, requested in order to repair other bikes. Finally, the bicycle in question is dismantled to such a degree that there is little hope of making it usable anymore. In order to prevent a defective bicycle from becoming such an object, some owners hide theirs in a storing room behind a locked door. However, even this drastic measure has limited effect, because others in the village will know that there is an unused bike with so many parts still in good condition and will come looking for it.[28]

There are also parts added to bicycles with the aim of expanding the range of possible usages. This is the case of the rear baggage carrier, already mentioned in the historical context from Nigeria. In Burkina Faso, baggage carriers are manufactured locally by blacksmiths. The local models are quite heavy (made from rods of structured steel), and very resilient. The ability to use the rear baggage carrier to transport heavy loads or as a second seat explains the preference of the owners for this local model. Typical loads in the rural area are sacks with cereals that are brought in from the fields, or agricultural tools. A plough on the back of a bicycle is a frequent sight, as are firewood and canisters of water or locally brewed beer.[29]

Another item added to the bicycles is the kickstand. Within two or three months after acquisition, most bicycles will no longer have their original stands. Many owners decide that the bicycle is fine without any kind of stand. However, those who buy a new stand tend to prefer the local model. Made by blacksmiths, it is heavier but more resistant than the original one. Other parts to be replaced are the rubber blocks of the pedals. Here again, the original pedals do not last long, and, consequently, many bicycle owners simply use the iron rods that were the former axles of the pedals. More durable pedals are made with rubber blocks, cut from car tires. Many bicycle tires are doubled from the interior with ribbons cut out of old and worn tires. It is not without irony that this is called "insurance." It protects the tube against holes, which occur quite frequently due to the bad road conditions.

Other modifications have little to do with functionality, but contribute to the appearance of the bicycle. Colored tape and plastic tubes are common decorations. The tape is attached to the handlebar and the frame; the tubes decorate the rods of the brakes. Further popular decorations are red, green, or yellow grips from soft rubber or foam material.[30] However, for older, intensely used bicycles, people will consider such decorations to be inappropriate, given the utilitarian context of the older bicycles' use.

Figure 1.1. Bicycle transport of ceramics.

The man is transporting a huge pot, made by his wife, to the market. Photo taken in 2001 in Northern Togo by the author.

Bicycle owners use old and new bikes differently. The difficult road conditions and heavy loads make it necessary to repair tubes frequently, change used tires, replace broken spokes, and weld rims or forks. By means of these frequent and sometimes complex repairs, bicycles achieve a particular appearance, and—paradoxically—only after these experiences and hardships will the owner really appreciate his bicycle. These events give him a better knowledge of his bicycle, convince him of its enduring character, and make him trust its resilience. This is not a question about the state of any single part, but rather a subjective evaluation of a bike after a longer usage, and the distances and itineraries managed with it. For these reasons, it is not the new bicycle that has the highest evaluation, but rather the well-worn bicycle, the one that is more or less reduced to its essential parts, and augmented by the addition of important local parts.

A comparison with a new bicycle may clarify the particular appreciation of the old. The new bike is labeled as the "bicycle of the white man"; each part and feature is associated with a functionality engineered in the factory. The local form, which represents what can be called the "Africanized bike," has lost some of the original functions as it has acquired others. However, the reduced number of functions is countered by a higher reliability: what is no longer part of the bicycle cannot be damaged. Though Africanized bicycles may need repair as well, the owner is familiar with and prepared to execute the necessary maintenance by himself or with the help of a "fitter." Although no bike user in West Africa would reject the idea that the bicycle of the white man is a kind of "optimal" bicycle, most bike users in West Africa appreciate the local bicycle's optimization with regard to local purposes and local possibilities of maintenance. The appropriated bicycles comply with the needs of the everyday uses and give the users the feeling of self-esteem, confidence, and trust.

Negotiating Contexts

The concept of cultural appropriation has been applied in the previous section in the domain of material transformation and incorporation. Ethnographic data have made clear that bicycles in use in West Africa differ materially to a considerable degree from the design of factory-new items. Furthermore, these changes have been explained by the particular environment and expectations of bicycle owners in Africa. A further aspect of cultural appropriation deals with the question of "objectification." Objectification explains how the socially accepted meanings of the bicycle emerge.[31]

The process of objectification is of general interest insofar as patterns of consumption are rapidly changing in Africa, and simultaneously the sphere of the social is redefined again and again. What constitutes appropriate forms of consumption requires an agreement on a socially accepted level, even if consumption practices are frequently contested. One widespread tendency is the disproportional increase of spending on clothes. Gerd Spittler has shown on the basis of a case study in Niger that such changes are also a consequence of changing hierarchies of values: food is devalued, and clothing becomes more important.[32] Many new patterns of consumption are related to the rising appreciation of urban lifestyles, associated with a regular income and sufficient pecuniary resources (not only self-produced food, but money). This is the way of life linked with the urban citizen, who is educated and employed by government or a large company.

Appreciation for new ways of living may also drive the acquisition of a bicycle in southern Burkina Faso. As already mentioned, the ownership of a bicycle is linked in most cases to a preceding period of labor migration. Therefore, owning a bicycle is associated with success in labor migration, the ability to work, and the ability to earn a regular income. There is a mutual relationship between bicycle ownership and social status: the bicycle is an object of conspicuous consumption.

These considerations about prestige and bicycles, however, are not a full description of the cultural appreciation of bicycles. The more practical aspects, as I have described above, are of at least equal importance. These vehicles are a prerequisite for the cultivation of the so-called bush fields, which are situated at a distance of 20 kilometers from the village or farther. The same applies for gardening. The harvest from the gardens—tomatoes, lettuce, or other vegetables—and commodities of local manufacture, such as beer, have to be brought to consumers or wholesalers, who are both in the major towns. Farmers need bicycles as affordable means of transport in order to generate a minimal income from bush fields or gardening. These are the only activities through which farmers can produce cash crops in larger quantities in order to amass the amount of money needed to buy, for example, a bicycle.[33] The lack of a bicycle may therefore constitute a vicious circle: the most profitable activities in the rural sector require the investment of large amounts of money and in particular the ownership of a bicycle. The bicycle is not just a prestige item: it is of equal importance for managing agriculture. Besides, the bicycle is a resource that changes many economic activities. For example, many women who brew beer use bicycles in order to transport the beer to the market.

Everybody in the village knows who owns a particular bicycle. The stories about the acquisition of most bicycles are public knowledge, as are their ages and brands. Awareness of and public interest in these facts are indicators of the relevance of bicycles. Therefore, bicycles count among the most valuable consumer goods in the households.[34] However, a bicycle is not an exceptional possession—this distinguishes bicycles from motorcycles. Aside from the teachers' motorcycles, there are no motorized vehicles in the village. The popularity of the bicycle finds its visual expression in any of the regular markets. In a corner of any market, one can find dozens of bicycles.

Although both men and women use bicycles, they use them for different professional activities. This differentiation concerns in particular the trip to the market, which is very often a professional activity for women, but a leisure trip for men. Bicycle usage is also a question of age, and may become an issue between different generations in a household. In this context, a juvenile migrant may claim the bicycle for himself, whereas others advise him to leave it for his father. The bicycle thereby may constitute a challenge for the authority of the elderly.

The villagers also critically evaluate ways in which people care for and about their bicycles. Those who show off their possession, invest too much in their bicycles, and misuse them for self-promotion are criticized with a proverbial saying: "This man has married his bicycle." In general terms, marrying is the basis of social relations for any adult in the society. Marrying a bicycle therefore is a fundamental criticism of someone who excessively uses the bike to flatter himself and disregards his social responsibilities.

This criticism addresses the more general question of how to deal with material possessions. Those who are accused of marrying a bicycle spend their money on decorations and spare parts and waste their time on useless repairs. Instead of taking social responsibility in a true marriage, or in useful labor, means and time are misdirected toward a material object. The proverb implies further social judgment if the aspects of exclusivity and commonality are considered. The rights in terms of bicycles include the explicit rules of lending and thereby sharing this resource with friends and neighbors. He who has married his bicycle forgets about these obligations and denies the social dimension of his possession. A third aspect refers to the gender of the bicycle. Bicycles are not exclusively male or female. Its gender is never something exclusive or evident, as should be the case in a proper marriage. Anthropomorphism with bicycles, or the identification with the opposite sex ("the bicycle is a woman"), is therefore criticized by the village's members.

To say that someone marries his bicycle makes a wrong behavior public. It points to unacceptable ways of dealing with a bicycle, and defines the local framework of cultural meanings of property. It is not a set of normative prescriptions, but merely a shared understanding about the limits of the "normal" when it comes to bicycles. The local meanings of bicycles are negotiated between the idea of prestige, associated with one of the most costly objects in the household, and the need for pragmatic usage in an environment with harsh economic and physical conditions.

Along its way from a global commodity to a local good, the "iron horse," as it is called in the area, is subject to a thorough transformation, materially and in terms of meanings. The usages of the bicycle in the village and its sometimes-contested ascription of usages and meanings furthermore reveal its impact on the transformation of the society.

Sustainability

It is obvious that bicycles in the contexts described here substitute motorized transports and thereby substantially contribute to sustainable models of transportation. Speaking in economic terms, people's use of bicycles is motivated by poverty; there are no other transport resources. With regard to the high prices for fuel, low yields for selling cash crops, difficult road conditions, and minimal benefits for retail traders, the bicycle is the only solution for the transportation requirements of farmers, pupils, traders, and many other villagers. It is not surprising that the agents of development have pointed to the potential role of bicycles and other nonmotorized means of transport (wheelbarrows, handcarts) for solving transport issues in an era beset by fossil-fuel pollution.[35]

However, it is worth noting that among the rural bicycle users who were the focus of this study in Burkina Faso, there is no reflection on the concept of sustainability. This lack leads to further considerations about the concept of sustainability as such. The case study reminds us that pathways to sustainability are not only by intentional design, selective innovation, and the strategic development of energy-saving devices. As development studies have demonstrated, local knowledge includes specific approaches to sustainability, and it can offer a critique of some assumptions about sustainability as innovation. The creative and improvised bicycle usage described here is not the outcome of a particular consciousness about the environment or the desire to practice a "new model" or a "sustainable innovation." Instead, the appropriation of the bicycle in Africa has been a site of the development and exercise of local knowledge.

This is the consensus of the available studies on bicycle use in Africa from countries such as Uganda, Malawi, and Ghana. They stress the fact that bicycles are valuable for making a living under harsh economic conditions and in the absence of any income-generating alternatives.[36] Bicycles are not the vehicles of choice, but available means to organize everyday life where other opportunities are out of reach. A differentiated evaluation of the bicycle in Africa needs to take into account the reality of poverty and its limited capacity to generate benefits. Sustainability here is not intentional, but evident through the practices on the ground.

There is no generally accepted opinion about which kind of bicycles might be the best for the conditions of use in Africa, nor are there any producers putting out a model exclusively tailored to African needs. Whereas some urge the creation of a local bicycle industry, the dominant practice is the importation of new and used bicycles from Europe and China.[37] However, the debate on importation or local production is missing a central point for a proper evaluation of bicycle usage in Africa. In answering the question of the "proper African bicycle," one would do well to consider the African cyclists' thorough transformation of their bicycles, as has been described here, to suit local needs and cultural contexts.

Indeed, the example of the "Africanized bicycle" may contribute to the redefinition of the concept of sustainability, as it has been argued in the introduction to this chapter. What stands out here is the multipurpose use of available bicycles and parts and the reduction of the bicycle's mechanic functionality to the essentials, with an eye toward efficiency, reliability, and longevity. Sustainability emerges not through an abstract plan, but through a thorough process of adaptation and transformation that converts bicycles into suitable devices of transport, for labor and leisure. Taking multipurpose functionality and reduced technical complexity as assets of sustainable technology, this example asks for a broader vision of the concept of sustainability.

Expanding on the reflections in this book's introduction, this chapter presents a somehow specific, but definitively very unconventional example of sustainability in the field of bicycle repair and extended usage. Rightfully, the introduction refers to the economic origins of the principles of sustainability three hundred years ago. It appears as if the practices described in this chapter are much closer to the original formulation of sustainability, despite bicycle repairers in West Africa never having heard about Hans Carl von Carlowitz. It would be a step in the direction to a more open, differentiated understanding of sustainable ways of life to consider the particular characteristics of sustainability in this case study not only as an outcome of explicit design, but also of improvisation and creativity.

Hans P. Hahn is professor of anthropology with a focus on African studies at the Institut für Ethnologie (Goethe-University, Frankfurt, Germany). His regional specialization is West Africa. His research profile includes material culture, consumption, and the impact of globalization on African societies. He recently edited a book, *Consumption in Africa* (2008), focusing on household economies in Africa. His most recent book, *Mobility, Meaning and Transformations of Things* (2013), deals with the more general aspects of artifacts that travel between continents, and thereby between cultures with the "mobility of things." He participated in a research program on local action and the impact of globalization in Africa (2000–2007). His publications include articles on the perception of the material world and an introductory book on material culture (2005). He recently published on the subjects of mobile phones in Burkina Faso, bicycles, and various other material items. He is the vice speaker of the research training group Value and Equivalency at Goethe-University.

Notes

1. Kwame Anthony Appiah, *In My Father's House* (New York, 1992), 157.
2. The author thanks the editors of this special section for integrating a contribution on Africa. He also expresses his gratitude to the anonymous reviewers who inspired substantial enhancements to this chapter.
3. John Howe, "'Filling the Middle': Uganda's Appropriate Transport Services," *Transport Reviews: A Transnational Transdisciplinary Journal* 23, no. 2 (2003): 161–176.
4. Jürgen Heyen-Perschon, "Das Fahrrad in Afrika – Luxus oder Notwendigkeit," *Praxis Geographie* 32 (2002): 16–20.
5. Hans P. Hahn, "Global Goods and the Process of Appropriation," in *Resistance and Expansion: Explorations of Local Vitality in Africa*, ed. Peter Probst and Gerd Spittler (Münster, 2004), 211–230.
6. Hans P. Hahn, "Globale Güter und lokales Handeln in Afrika," *Sociologus (N.F.)*, no. 54 (2004): 1–23.
7. This notion of modernity is basically that of Jonathan Friedman's; see Friedman, "Modernity and Other Traditions," in *Critically Modern: Alternatives, Alterities, Anthropologies*, ed. Bruce M. Knauft (Bloomington, IN, 2002), 287–313. Friedman's arguments are in line with several other, partially more recent studies that show how modernity in Africa has a specific evaluation, critically adopting aspects of "standard assumptions" about modernity—see Jean Comaroff and John L. Comaroff, *Modernity and its Malcontents: Ritual and Power in Postcolonial Africa* (Chicago, IL, 1993); and Charles D. Piot, "Of Hybridity, Modernity, and Their Malcontents," *Interventions* no. 3 (2001): 85–91—but also self-consciously redefining modernity in a context of despair and poverty.
8. Compared to the bicycle, the concern for understanding the role of cars and motorbikes is much more important. Cf. Jan-Bart Gewald, Sabine Luning, and Klaas van Walraven, eds., *The Speed of Change: Motor Vehicles and People in Africa, 1890–2000* (Leiden, 2009); Piet Konings, "'Bendskin' Drivers in Douala's New Bell Neighbourhood: Masters of the Road and the City," in *Crisis and Creativity. Exploring the Wealth of the African Neighbourhood*, ed. Piet Konings and Dick Foeken (Leiden, 2006), 46–65; Joost Beuving, "Cotonou's Klondike: African Traders and Second-hand Car Markets in Benin," *Journal of Modern African Studies* no. 42 (2004): 511–537. Xavier Godard and H. Ngabmen, "Comme Zemidjans, ou le succès des Taxi-Motos," in *Les transports et la ville en Afrique au sud du Sahara. Le temps de la débrouille et du désordre inventif*, ed. Xavier Godard (Paris, 2002), 397–406.
9. John D. N. Riverson and Steve Carapetis, *Intermediate Means of Transport in Sub-Saharan Africa: Its Potential for Improving Rural Travel and Transport*, World Bank Technical Paper 161 (Washington, DC, 1991); Jonathan Dawson and Ian Barwell, *Roads Are Not Enough: New Perspectives on Rural Transport Planning in Developing Countries* (London, 1993); Christina Malmberg Calvo, *Case Studies on Intermediate Means of Transport Bicycles and Rural Women in Uganda*, SSATP Working Paper 12 (Washington, DC, 1994); The World Bank, *Sustainable Transport: Priorities for Policy Reform*, Development in Practice Series (Washington, DC, 1994).
10. See Peter Cox, *Moving People: Sustainable Transport Development* (London, 2008).
11. Cf. Gina Porter "Living in a Walking World: Rural Mobility and Social Equity Issues in Sub-Saharan Africa," *World Development* 30, no. 2 (2002): 285–300.

12. There are quite a few exceptions to this, dealing with the hidden popularity of bicycles among the local population. All reports point to bicycles being more useful than acknowledged by colonial government. Lucie Mair, in *An African People in the Twentieth Century* (London, 1934), reports on people going regularly from their rural homesteads to their urban workplaces on bicycles. Another article—Luther Gerlach, "Traders on Bicycles: A Study of Entrepreneurship and Culture Change among the Digo and Duruma of Kenya," *Sociologus (N.F.)* 13 (1963): 32–49—refers to the utility of bicycles for regional trade in Kenya. A third case study from Richard Thurnwald, *Black and White in East Africa* (London, 1935), esp. 322ff., contains a more general appraisal of bicycles in Tanzania. Nancy Rose Hunt, in "Bicycles, Birth Certificates, and Clysters: Colonial Objects as Reproductive Debris in Mobutu's Zaire," in *Commodification: Things, Agency, and Identities*, ed. Wim van Binsbergen (Münster, 2005), 123–141, reports about the usage of bicycles by nurses in colonial Congo.
13. Engaging the carriers was a major problem for many travelers. One example of such an expedition, which suffered from enforced halts due to the lack of carriers, comes from Leo Frobenius, who was in the same area as Fisch at roughly the same time; Frobenius, *Auf dem Wege nach Atlantis ... Bericht über den Verlauf der zweiten Reiseperiode der DIAFE in den Jahren 1909/1910* (Berlin, 1911).
14. Rudolf Fisch, *Nord-Togo und seine westliche Nachbarschaft*, trans. H. P. Hahn (Basel, 1911), 113.
15. Hans Schomburgk, *Bwakukama: Fahrten und Forschungen mit Büchse und Film im unbekannten Afrika*, trans. H. P. Hahn (Berlin, 1922), 125.
16. That colonial administrators sometimes had shown a different attitude toward bicycles is documented by Jan-Bart Gewald, "People, Mines and Cars: Towards a Revision of Zambian History, 1890–1930," in *The Speed of Change*, ed. Gewald, Luning, and van Walraven, 21–47, here: 36. The travelogues from 1901 quoted by Gewald show a considerable disdain for the usage of bicycles in Zambia.
17. This shift from Europeans to Africans as bicycle users is also reported by Terence Ranger, "Bicycles and the Social History of Bulawayo," in *Short Writings from Bulawayo*, ed. Jane Morris (Bulawayo, 2003), 76–81. Bicycle races of Africans and the widespread use of bicycles by Africans in Congo before 1914 have been documented by Nancy Rose Hunt, "Nurses and Bicycles," in *A Colonial Lexicon of Birth Ritual, Medicalization and Mobility in the Congo*, ed. Rose Hunt (Durham, NC, 1999), 159–195.
18. This is also confirmed by Rex U. Ugorji and Nnennaya Achinivu, "The Significance of Bicycles in a Nigerian Village," *The Journal of Social Psychology* 102 (1977): 241–246. These authors report on the moment in 1912 when the first bicycles arrived in Umoru, an Igbo village.
19. This is documented by Anthony I. Nwabughuogu, "The Role of Bicycle Transport in the Economic Development of Eastern Nigeria, 1930–45," *Journal of Transport History*, Third Series 5 (1984): 91–98.
20. For more information about bicycle industries in China, and the Phoenix brand in particular, cf. Amir Moghaddass Esfehani, "The Bicycle's Long Way to China: The Appropriation of Cycling as a Foreign Cultural Technique 1860–1940," in *Cycle History 13. Proceedings, 13th International Cycling History Conference*, ed. Andrew Ritchie and Rob van der Plas (San Francisco, CA, 2003), 94–102.

21. To see this advertisement consult: http://farm1.static.flickr.com/113/262119954_f3 72b153f6.jpg (accessed 8 September 2014). Similar advertisements are reported from the 1950s by Ranger, "Bicycles," 79. Cf. also Gregory H. Bowden, *The Story of the Raleigh Cycle* (London, 1975), 72, 73.
22. Bowden, *The Story of the Raleigh Cycle*, 139.
23. Amir Moghaddass Esfehani (personal communication) found advertisements from Nigerian tradespeople in the journal "Shangye Yuebao" (*Business Monthly*, Shanghai) from 1946. These advertisements called for Chinese partners for dealings with bicycles and bicycle parts.
24. David K. Fieldhouse, *Merchant Capital and Economic Decolonization: The United Africa Company, 1929-1987* (Oxford, 1994).
25. The larger framework of this research was a focus on consumption patterns in a rural setting and the impact of "globalization." Part of this research was an investigation on material possessions of rural households and the perception of global goods by both men and women; Hans P. Hahn, Gerd Spittler, and Markus Verne, "How Many Things Does Man Need? Material Possessions and Consumption in Three West African Villages," in *Consumption in Africa*, ed. Hans P. Hahn (Münster, 2008), 115-140.
26. Jojada Verrips and Birgit Meyer, "Kwaku's Car: The Struggles and Stories of a Ghanaian Long-distance Taxi-driver," in *Car Cultures*, ed. Daniel Miller (Oxford, 2001), 153-184.
27. Kurt Beck, "Die Aneignung der Maschine," in *New Heimat*, ed. Karl-Heinz Kohl (New York, 2001), 66-77.
28. Brad Weiss reports on a similar context in Tanzania: old bicycle frames are stored in some households just because there was once the expectation to remount all missing parts; Weiss "Forgetting your Dead: Alienable and Inalienable Objects in Northwest Tanzania," *Anthropological Quarterly* 70, no. 4 (1997): 164-172.
29. More examples with modified back carriers from Ghana and Uganda are shown in Barbara Bloemink and Cynthia Smith, *Design for the Other 90%* (Washington, DC, 2007), esp. 2 and 87.
30. Reports from Nigeria and Zambia on such decorations emphasize the skillfulness and creativity in selecting colors and motives; Elisha P. Renne and Dakyes S. Usman, "Bicycle Decoration and Everyday Aesthetics in Northern Nigeria," *African Arts* 32, no. 2 (1999): 46-51. See also Yvonne Vera, "Thatha Bhasikili: Adorned Bicycles in Bulawayo," *Gallery: The Art Magazine from Gallery Delta* (September, 2001): 8-10.
31. Considering appropriation as a process—including objectification, incorporation, and contextualization—is a concept developed in the framework of the author's research; Hans P. Hahn, *Materielle Kultur: Eine Einführung* (Berlin, 2005), esp. 102. However, similar notions of appropriation, distinguishing between incorporation and contextualization, have been applied earlier in media studies and also in the study of cars; Eamonn Carrabine and Brian Longhurst, "Consuming the Car: Anticipation, Use and Meaning in Contemporary Youth Culture," *The Sociological Review* 50, no. 2 (2002): 181-196.
32. Gerd Spittler, "Kleidung statt Essen. Der Übergang von der Subsistenzproduktion zur Marktproduktion bei den Hausa (Niger)," in *Afrika zwischen Subsistenzökonomie und Imperialismus*, ed. Georg Elwert and Roland Fett (Frankfurt, 1982), 93-105.

33. Ursula Barth, *Frauen gehen lange Wege. Transportvorgänge von Frauen in ländlichen Regionen Afrikas* (Karlsruhe, 1989); and Ian Barwell, *Le transport et le village. Conclusions d'une série d'enquêtes-villages et d'études de cas réalisées en Afrique* (Washington, DC, 1998).
34. This is also the reason why in colonial times there was a tendency to charge bicycle owners a special bicycle tax. This was the case for example in Mali: Torild Löfwander, *Die sozialökonomischen Verhältnisse der bäuerlichen Bevölkerung in der Republik Mali* (Berlin, 1983).
35. Jimu hails the contribution of these means of transport for the economy in Malawi. However, most people earning their living with any of these means of transportation do this for the lack of any alternative. Bicycles in Malawi generate a limited income, but they do not provide wealth for these hard-working people; Igansio M. Jimu, *Urban Appropriation and Transformation: Bicycle Taxi and Handcart Operators in Mzuzu, Malawi* (Mankon, 2008).
36. The World Bank, *Sustainable Transport—Priorities in Practice* (Washington, DC, 1996), esp. 76; and Jürgen Heyen-Perschon, *Nichtmotorisierte Transportmittel und ihr Einfluss auf die wirtschaftliche und soziale Entwicklung in ländlichen Räumen Ugandas. Empirische Untersuchung zur Kosten-Nutzen-Effizienz der Fahrradnutzung* (Hamburg, 2002).
37. See http://web490.server-drome.net/bestbike.html (accessed 8 September 2014) for a discussion of this problem. For the difficult experience of designing and importing "special" bicycles, cf. Bradley Schroeder, "Doing Business in Africa: The California Bike Coalition Comes of Age," *Sustainable Transport* 18 (2007): 18–21, 30.

 CHAPTER 2

The Politics of Bicycle Innovation
Comparing the American and Dutch Human-Powered Vehicle Movements, 1970s–Present

Manuel Stoffers

The front cover of the March 1973 issue of the popular science magazine *Scientific American* featured a painting of a racing bicycle, indicating the topic of the main article: bicycle technology. In the ten-page essay, author Stuart S. Wilson, a lecturer in engineering at the University of Oxford, presented an elaborate technological history and appraisal of the bicycle as "the most benevolent of machines," stressing its unsurpassed energy efficiency compared not only to all other vehicles but also to all "moving creatures," and recommending its increased use in both the "underdeveloped" and "overdeveloped" world. "If one were to give a short prescription for dealing rationally with the world's problems of development, transportation, health and the efficient use of resources," Wilson concluded his eulogy, "one could do worse than the simple formula: Cycle and recycle."[1]

Wilson's article was indicative of a change in the evaluation of the bicycle in Western countries during and after the 1960s and 1970s, a period in which a range of new social movements started to advocate the increased use of the bicycle for transportation. In many Western countries, new bicycle advocacy organizations, often of environmentalist inspiration, came into being, while some older cyclists' interest organizations were revived, and new forms of cycling advocacy, such as Critical Mass (since 1992) and the World Naked Bike Ride (since 2004), spread around the Western world. Urban visionaries and planners started to rethink the place and potential of cycling within the urban traffic system.[2]

"Reframed" by all these initiatives, the bicycle has now ceased to be an obsolete vehicle for urban transportation—at least in the minds of trendsetting young urban elites and in the mobility discourses of many policy makers in the Western world.[3] To cut short a long list of documentation in support of this statement: since the 1990s, even the US government has introduced legislation to encourage cycling as a mode of transport, heralding according to some "the modern era of biking."[4] To whatever extent this statement needs to be qualified—after all, to anyone reflecting on forty years of insistent bicycle activism, the net result in many countries may seem insubstantial—there is no doubt that the bicycle has indeed become *modern* again, for the second time in its history.

This essay focuses on a single aspect of this remarkable turn in the Western appreciation of cycling, which captures the reinvention of the bicycle perhaps most concretely and surprisingly: the spread of the human-powered vehicle movement at the end of the twentieth century. Previous histories of this small-scale international movement of engineers and scientists who started to radically rethink bicycle technology were mostly written by people who were directly involved in the movement and who focused on its technological or record-breaking achievements.[5] Other publications on the topic have addressed the influence of restrictive sports regulations on bicycle innovation, have demonstrated problems of new product development, or have reflected on possible contributions of innovative human-powered vehicles to sustainable mobility.[6] Instead, the present contribution will highlight the relevance of political ideals and national cultures of cycling to understand the origins and the development of these organized, collective, and noncommercial attempts to innovate the bicycle, which were an integral part of late twentieth-century efforts to reconcile modernity with sustainable lifestyles.

Founded in 1976, the International Human-Powered Vehicle Association (IHPVA), inspired by the idea to promote cycling through technological innovation and scientific research, was both the ultimate expression and exponent of the growing opposition against the view of the bicycle as an outdated and backward vehicle in the West. At least partly driven by environmentalist concerns, the IHPVA aimed at improving the traditional bicycle, striving to make it a closer competitor and attractive alternative to motorized vehicles as a mode of transportation. The sort of vehicle that the engineers of the IHPVA had in mind could be seen on another *Scientific American* cover, published ten years after Wilson's praise for the unequalled efficiency of the traditional bicycle: a futuristic, streamlined, human-powered tricycle—the tandem-version of which had just surpassed the speed threshold of 100 km/h in 1980.[7]

The human-powered vehicle movement was part of the broader "alternative technology" movement that broke away from the Neo-Luddite tendencies that characterized other counterculture and environmentalist movements at the time.[8] Best exemplified perhaps by Stewart Brand's well-known *Whole Earth Catalog*, the alternative-technology school was convinced that clever technology could contribute to a sustainable and just world and solve pressing problems of environmental degradation and energy resource depletion. In the words of historian Andrew Kirk, "unbridled technological optimism" was one of its distinctive—and, for many, attractive—characteristics.[9] As

Figure 2.1. An HPV utopia: human-powered vehicles on the road, on the water, and in the air.

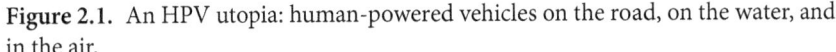

Painting by Greg Trayling, originally designed to announce the IHPVA Speed Championships held at the EXPO 86 World Exposition on Transportation and Communication in Vancouver, Canada (May–October 1986). During one particular week, EXPO 86, in close cooperation with the IHPVA, paid special attention to "Human-Powered Transportation." Reproduced with kind permission from Greg Trayling.

will be seen, the same characteristics applied to the human-powered vehicle movement.

Originating in the United States in the middle of the 1970s with an explicitly international outlook, the human-powered vehicle movement soon spread to Europe and other parts of the Western world, including the Netherlands. One might expect that the HPV movement, collectively working to improve the bicycle, would have met with major success in a cycling country such as the Netherlands. However, the reality turned out to be different. Whereas in the United States, the IHPVA tried to adjust the image (and technology) of cycling to meet the dominant automotive ideals of speed, comfort, and safety, in "the land of bicycles," Dutch HPV enthusiasts faced the problem of having to compete with a strong bicycle culture that was neither geared toward speed, comfort, or safety issues, nor toward innovation. In fact, despite its international orientation and ethos, national varieties in practices and ideas of the human-powered vehicle clubs quickly developed.

Based on a reading of magazines and other publications from the American and Dutch HPV scenes, this essay compares and contrasts ideas and practices within the Dutch human-powered vehicle association NVHPV (est. 1984) with those of the American-based IHPVA, thus contributing to our understanding of the cultural complexities of the Western bicycle renaissance of the last few decades and the ideas to promote sustainable transportation that were connected with it.[10] In the end, one realizes that while ideas to promote cycling may travel easily, national trajectories and cultures of cycling prove remarkably resilient to change.

The IHPVA: A New Social Movement Devoted to Bicycle Innovation

The founding of the IHPVA directly followed the high tide of the Great American Bicycle (Sales) Boom of 1972–1974, in which the number of adult bicycles sold increased by a factor of twenty.[11] It also followed soon after fuel shortage, caused by the 1973–1974 OPEC Oil Embargo against the United States, had led to short-lived but widely publicized commercial experiments with "pedal-powered cars."[12] The IHPVA's cradle stood in California, a center of the alternative youth culture that soon after also gave birth to the mountain bicycle.[13] Here, in Seal Beach, Chester Kyle, professor of mechanical engineering at nearby California State University, and a few other engineering enthusiasts, founded the International Human-Powered Vehicle Association on 28 March 1976.[14] Its first and foremost purpose was to organize and regulate races between prototypes of aerodynamically improved, human-powered bicycles and tricycles (as well as between human-powered boats and aircraft). Back in 1973, Kyle, together with a few of his students, had started to experiment with the aerodynamics of bicycles, and in 1975 he had organized the first of a still-continuing series of International Human-Powered Speed Championships.[15] As the innovative and experimental human-powered vehicles in these competitions did not fit the rules of existing bicycle racing bodies, a new organization was needed to recognize and document the remarkable speed records that were achieved—and get them published in the *Guinness Book of World Records*.[16]

While speed records thus were without doubt IHPVA's core business, racing was not an end in itself. Looking back at the beginning of the movement just a few years before, the first issue of *Human Power*, the magazine of the IHPVA, stated in 1977: "It was hoped that the stimulus provided by Speed Championships and by other events sponsored by the IHPVA would encourage radical changes and improvements in human-powered transportation throughout the world."[17] In a contribution to the popular science and science-fiction magazine *Omni*, Kyle predicted in 1979 that on the day that Western oil supplies would be exhausted, "there may be a beautiful, streamlined bicycle waiting for us that anyone in reasonable condition will be able to pedal along at commuter traffic speeds, or even faster."[18] Within the IHPVA, the most outspoken early representative of these broader ideas behind the HPV movement was David Gordon Wilson, another professor of mechanical engineering, working at MIT (and not to be confused with his Oxford colleague Stuart S. Wilson). Wilson became involved in the IHPVA from the very beginning, first as a member of the board of directors, then as its seventh president (1983–1984), and subsequently as the long-time editor-in-chief of *Human Power*. In fact, Wilson had been instrumental in preparing the ground from which the IHPVA could emerge. In 1967, he had already organized a widely publicized international design competition to innovate bicycles.[19] In 1974, he coauthored *Bicycling Science* (1974)—the only English-language textbook on cycling physics, addressed to an international academic audience, and in its third edition still available from MIT Press. In it, Wilson had broadened the scope from (traditional) bicycles to "human-powered machines" in general and addressed the need and possibilities for human-powered transportation in the future.[20]

Although Wilson, like Kyle, acknowledged aerodynamic drag to be, quite literally, a drawback of the traditional bicycle, he focused more on improving safety and comfort than on increasing the speed potential of bicycles. Whereas the first issues of *Human Power* only showed very fast, but also very impractical, streamlined record machines, Wilson's articles in these years included reflections upon an alternative type of bicycle for daily use, the "recumbent bicycle" (or "recumbent" for short): a low, long-stretched, and feet-first type of bicycle with a seat instead of saddle, but not necessarily with an aerodynamic hull. Similar bicycles had been produced before the Second World War, but were now being reinvented and rediscovered.[21] Wilson's experiments in the early 1970s, in cooperation with Californian builder Fred Willkie, finally resulted in one of the first of many modern recumbents to appear on the bicycle market in the last decades of the twentieth century: the Avatar 2000, a long wheelbase recumbent bicycle with under-seat steering that became commercially available in early 1981 and subsequently would inspire many HPV builders.[22]

More explicitly than Kyle, Wilson emphasized the need for what he called an "HPV revolution" that would lead to a "major increase in the use of human-powered vehicles (HPVs) for commuting, shopping and recreation."[23] In 1967, the inspiration for his bicycle design contest had come from concerns over the urban pollution, congestion, and "unhumanity" caused by the ever-increasing use of the car.[24] The early Wilson-Willkie recumbent designs were, significantly, called "Green Planet Specials," and in *Bicycling Science* Wilson warned against "irreversible changes in the whole of earth's ecology."[25] Thirty years later similar ideas were still alive within the IHPVA, as is shown by the following quote from IHPVA chair and internationally bestselling bicycle author Richard Ballantine, written shortly after the beginning of the Iraq War of 2003.[26]

> The politics of HPVs are simple. In developed countries, HPVs can meet a large proportion of transport requirements, and in so doing, reduce consumption of petroleum. The decrease in size of environmental footprint in terms of resource depletion means less incentive for taking oil through economic and military force. As well, there is less damage to the environment and human health from pollution. Finally, and not least, HPVs suit modern demographic trends. The population of the world is increasing, and at the same time, more and more people are living in cities. In high density urban environments, where space is limited, bikes and HPVs are not just hugely more energy-efficient than motor vehicles, they are also faster. Bombs and cars go together—and in the end, HPVs will beat both.

Bicycle innovation was the IHPVA's answer to countering the adverse effects of increasing automobility. Although both Kyle and Wilson acknowledged that bicycle innovation was not enough to make "a substantial proportion of the population freely [choose] HPVs as their means of commuting," the contribution of the IHPVA to this general aim was limited to efforts to improve the bicycle.[27] Typically, in a 1979 article, Kyle expressed his skepticism about the prospects for car tax schemes and separate bicycle infrastructure in the United States, while insisting on the practical relevance that the contraptions of the "irrepressible inventors" participating in the Speed Championships could acquire.[28] In 1982, such considerations led to the first of a series of "practical vehicle design competitions" organized by the IHPVA, in addition to the speed contests.[29]

Confidence in science and technology as instruments to solve social problems came naturally to the US-based HPV enthusiasts, many of whom appear to have had a technical or scientific background. Typically, of the twelve founders of the IHPVA, more than half were engineers, three had completed a dissertation, and two were employed at a university.[30] Many, if not most, contributions to *Human Power* were of a technical or scientific nature, and in the 1980s and 1990s, the IHPVA made the effort to organize several scientific symposia.[31] In 1995, leading members of the IHPVA published a collaborative academic volume on human-powered vehicles, presenting the latest knowledge on the development and propulsion of HPVs.[32] The confidence in technological progress was aptly reflected in the widespread belief among HPV enthusiasts in the superiority of their vehicles compared to traditional bicycles, but also in the popularity of utopian images in which a variety of HPVs—including flying machines—were seen to dominate the urban landscapes of the future. Especially in the first decades of its existence, the HPV movement attracted much public attention through futuristic, quasi-hi-tech HPV designs that seemed to promise clean yet rapid mobility by human power only—and made great magazine covers as well.[33] While being at least partly environmentalist by inspiration, HPV enthusiasts were certainly not environmentalists of the Neo-Luddite or "antimodernist" variety. Rather, they belonged to the alternative or appropriate technology school, as represented by Stewart Brand's *Whole Earth Catalog*, believing in the far-reaching solutions that simple yet inventive technology could bring.[34]

To sum up, the main activities within the IHPVA were developing HPV prototypes, organizing competitions between these vehicles, and sharing knowledge about them. The competitions were organized to encourage technical innovation and at the same time generate publicity for the bicycle-innovation cause.[35] HPV engineers saw it as their main technical challenge to make bicycles faster, but they also wanted to make bicycles safer and more comfortable, in particular by providing weather protection and more comfortable seats. These attempts to innovate the bicycle were prompted by concerns about the environment, health, and urban life. HPV enthusiasts shared these concerns with many other new bicycle advocacy and environmentalist movements that sprung up in the 1970s and 1980s. But the IHPVA's unique selling point was its confidence in a scientific approach to cycling and in the ability of innovative bicycle technology to contribute to the coveted increase of bicycle use in the future.

More specifically, many IHPVA members believed that innovative technology could make the bicycle a closer competitor to the car. In this respect, the IHPVA conformed to the "unique, indigenous American style" of bicycle advocacy known as "vehicular cycling": the idea that the interests of cyclists are best served *not* by creating separate facilities for cyclists (thus discriminating between cyclists and car drivers), but on the contrary by identifying cyclists as a sort of car driver, with equal ways, rights, and needs.[36]

Although there had been an American "sidepath movement" back in the 1890s, and demands for cycle paths were raised again by the cycling minority from the 1960s onward, the predominant and effective resistance to building separate cycling infrastructure received support even within the cycling community from the proponents of "vehicular cycling."[37] The best-known of them, John Forester, argued in his handbook for policy makers, *Bicycle Transportation* (1983, first edition 1977), that as far as the infrastructure was concerned, the United States was "the world's best nation for cycling, just as it is for motoring"—if only cyclists received the same treatment as car drivers, and started to behave like them.[38] In a remarkably similar way, the IHPVA's mission and strategy was to minimize the differences between bicycles/HPVs and cars. As Kyle put it in 1979: "Car lovers say—and this group includes the majority of the American people—the bicycle is *slow*, and when it rains you get wet. Well, that's simply no longer true."[39]

The International Beginnings of the Dutch HPV Club

Although the IHPVA originated in California, and Californians played an important role in the organization, for example by providing seven of the first ten presidents between 1976 and 1991, the IHPVA understood itself as an internationally oriented organization, as the chosen names of the Association and of the International Human-Powered Speed Championships well illustrate.[40] Soon HPV clubs were founded in other parts of the United States, and through an international board of directors, which included members from a number of European countries, Australia, and Japan, contacts on other continents were established from the very beginning.[41]

Publications on HPVs in books and magazines with an international readership, such as Wilson's *Bicycling Science* (1974, 1982), Ballantine's *Richard's Bicycle Book* (many editions and translations since 1972),[42] *National Geographic* (first article on HPVs in 1977), and *Scientific American* (first article on HPVs in 1983), spread the news about the American efforts to innovate the bicycle, the remarkable speed records, and the ideas behind them. Even behind the Iron Curtain, these developments did not go unnoticed, if only because the popular US propaganda magazine *Amerika*, appearing in Russian, chose to cover the HPV speed championships in 1977.[43]

Soon contests in imitation of the American HPV speed championships were organized elsewhere, for instance in Japan in 1978—after a television broadcast on American HPV speed championships—in Brighton (UK) in September 1980 and behind the Iron Curtain in Šiauliai (Lithuanian SSR) in 1982.[44] In several European countries, HPV clubs were established after the first, usually well-publicized, HPV races were held there: first in Britain (1983), then elsewhere, e.g., in Switzerland (informally 1984, formally 1985) and (West) Germany (informally 1985, formally in 1986), followed in the 1990s by France (1991), Denmark (1993), Belgium (1993), and several other countries.

The Netherlands belonged to the first wave of countries in Europe to organize HPV races and establish an HPV club, second only after the United Kingdom. It is worth noting the foreign, especially American, connections in the establishment of the Dutch HPV scene. Apart from a show race on Queen's Day in The Hague earlier that year, the first serious Dutch HPV speed contest was organized at the beach resort town of Zandvoort in September 1983 by "Holland's first recreational cycling magazine," *Fiets*. *Fiets* had been launched only the year before by American publisher Bob Rubinstein, who, while living in Amsterdam, was not burdened with the country's prevailing skepticism that the Dutch were not fascinated enough by bicycles to *read* about them.[45] From the very beginning, *Fiets* reported enthusiastically both about alternative developments in bicycle technology and international HPV races, then organized the first Dutch HPV races, and subsequently several members of the magazine staff became involved in the founding and organization of the NVHPV, the Nederlandse Vereniging voor Human-Powered Vehicles (est. 29 September 1984).[46] It was American *Fiets* staff member Cary Peterson who suggested the Anglo-Dutch name at the inaugural meeting.[47] Another founding member was Briton Ian Borwell, who had been involved in the HPV scene in the United Kingdom—co-designing with Mike Burrows the Windcheetah, a racy recumbent tricycle that became a classic of HPV design—before he moved to the Netherlands in 1982.[48]

Considering its international inspiration, it is not surprising that ideas and practices within the NVHPV were, at least initially, quite similar to those of the IHPVA.[49] The 1985 statutes of the Dutch association described its purpose in a similar way as the IHPVA statutes: "to promote the use and development of land, water and air vehicles, powered by human muscle."[50] To reach the goal, the NVHPV planned to organize competitions and meetings and publish a magazine. As the early volumes of the two-monthly club magazine *HPV Nieuws* show, the main activities of the NVHPV during the 1980s were organizing competitions between different HPVs and sharing information on technical challenges of newly built DIY recumbents—both through the club magazine and through the organization of "technical days" by the "technical activities commission" of the club.

Judging from the early establishment and growth of the Dutch HPV club, the HPV movement found fertile ground in the Netherlands. Membership of the NVHPV rose steadily from sixty in 1985 to one thousand members in 1993, then leveled down somewhat, but increased sharply again from 1997 onward to 2002, reaching a maximum of nearly two thousand members. By then, the Dutch branch was the biggest national HPV club in the world.[51]

Successful also, at least to some extent, were Dutch recumbent firms, the owners of which usually were at one time (active) members of the NVHPV. From spring 1983 onward, when the first commercial Dutch recumbent, the Roulandt, appeared on the market, recumbents were produced in increasing numbers by Dutch HPV builders who started a production firm—some of which still exist today: in the 1980s, M5 (production starting between 1983 and 1985), Jouta (1984), and Flevobike (1988), followed in the 1990s by Challenge (production starting in 1991), Sinner (1993), Optima (1994), Alligt (1995), Nazca (1999), J&S Fietsdiensten/Velomobiel (1999), Rainbow (1999), and others.[52] These small Dutch recumbent firms dominated the national market for recumbent bicycles. Although established Dutch bicycle firms initially followed the developments with some interest, and some incidentally experimented with recumbent production, in general they were very reluctant to add recumbent models to their product range.[53]

The increase in the number of recumbent riders, made possible by growing commercial production, soon became apparent in the new activities of the NVHPV, next to racing and building prototype HPVs. In 1989, NVHPV chairman Cor Moritz urged members to organize tours, in addition to "the existing race formula."[54] From that moment on, the club incidentally organized tours; at the same time, articles on recumbent tours and travels, but also on touring equipment and accessories, increasingly appeared in *HPV Nieuws*. Then, at the end of the 1990s, as the number of recumbent riders had increased further, local and regional recumbent clubs were formed, mainly to undertake bicycle trips together, and in 2000, the NVHPV created a Camping and Touring Commission that organized clubwide national touring events for recumbent cyclists and velomobile riders on a regular basis.[55]

Further opening up to the general public, the Dutch HPV club began in 1997 to organize Cycle Vision, a yearly public event that was not confined to racing; it included a fair, a second-hand market, and a testing ground for commercial recumbents. In the second half of the 1990s, the website of the club became a new important way of binding members and opening up to the public, providing daily news of general and detailed nature on recumbents, and sharing experiences—many pages being translated into more than one foreign language. At one point, the NVHPV also managed to receive funds from the major governmental bicycle promotion program *Masterplan Fiets* (1990–1997) to organize an HPV design contest.[56] Although in general, the Dutch government held that bicycle innovation should be left to the market, in the 1990s, public money was invested in the Mitka project, meant to develop an attractive, innovative human-powered tricycle that could lure car drivers out of their vehicles.[57]

The Deviating Profile of the Dutch HPV Club

While the reception of international HPV ideas thus led to a flourishing HPV club in the Netherlands, over time remarkable differences between the Dutch and American movements became apparent. The attention of the Dutch club magazine for technical experiments dwindled as soon as commercial recumbents became available. The involvement of academics was also decidedly less than in the IHPVA. Most remarkable, however, was the apparent weakness of the environmentalist, utopian, and idealist inspiration within the NVHPV, compared to its American counterpart, and the different direction its bicycle activism took.

Although the first issue of the Dutch club magazine *HPV Nieuws* called recumbents "one of the best alternatives for the car," the idea that bicycle innovation could help to promote bicycle use was not a very prominent one in the Dutch club.[58] Political considerations or bicycle activism did not play a prominent role for years. When the environmentalist organization in favor of alternative technology, Small Earth (*De Kleine Aarde*), asked NVHPV members in 1984 to participate in a seminar on the bicycle in the future "meant to discuss and present new developments in cycling—with a clear focus on recumbents," the NVHPV greeted the invitation with little enthusiasm. According to *De Kleine Aarde,* the NVHPV was too introverted, "focusing on developing new ideas individually and organizing competitive events," without caring much about the problem of "public acceptance."[59] Unlike the IHPVA, the Dutch club did not have prominent members such as David Wilson or Richard Ballantine, who wrote explicitly, elaborately, and inspiringly about the broader, social motivations behind bicycle innovation. Whereas the Swiss and German HPV clubs, which were also founded in the early to mid-1980s, explicitly added environmentalist aims to their official statutes, the Dutch club did not.

Still, the politics of bicycle innovation were not entirely absent. In 1989, the NVHPV put the development of a "365-day bicycle" on the club's agenda. Similar to American HPV thinking, such a "bicycle for all seasons" was meant to "serve as a replacement of the car," as the first major contribution on this topic in *HPV Nieuws* stated.[60] In the same year, the NVHPV installed a commission to push the idea. It took four years for the project to result in what became a successful bicycle design contest, in cooperation with *Fiets* in 1993. Not only did the contest receive a lot of media attention, it was also the first (and only) time the NVHPV was able to involve scientists in a project, inviting IHPVA-founder Chester Kyle and Dutch physiologist Van Ingen Schenau (next to the British HPV designer Mike Burrows) to participate in a small symposium at the Technical University Eindhoven.[61] Eventually, the prize-winning vehicle, the Alleweder ("Allweather"), a faired aluminum tricycle by manufacturer Flevobike, established the "velomobile" (i.e., a faired recumbent tricycle) as the ultimate commuter vehicle in Dutch HPV circles.

Subsequent initiatives along HPV lines met with less success, however. In 1995, the NVHPV thought of organizing another symposium on alternative vehicle concepts for a sustainable world, parallel to the planned HPV World Championships in Eindhoven, but this came to nothing after the championships were canceled.[62] At the end of the 1990s, the NVHPV again coorganized the *Bike 2000 Construction Contest* (1999), a design competition that was explicitly meant to help solve contemporary mobility issues.[63] The contest ended in disillusion, however. No entry was considered to meet the set design requirements: a lightweight, low maintenance bicycle, with a large degree of comfort and weather protection, very agile in the city, while aerodynamic in the countryside, and with a waterproof luggage compartment. Since then, nobody within the NVHPV has again dared to organize or even propose such a challenging project, for which apparently not enough enthusiasm or input could be mobilized.

That is not so say that bicycle activism played no role whatsoever in the Dutch HPV club. However, it moved away from the IHPVA approach of designing HPVs that could compete with cars. Instead, the bicycle activism of the NVHPV took another, typically Dutch and *nonvehicular* turn: the demand for a segregated cycling infrastructure, in this case specifically designed for fast moving recumbent bicycles and velomobiles. In the course of the 1990s, the initial emphasis within the NVHPV on speed contests not only gave way to touring, but also to utility cycling, especially commuting. In 1993, for example, an article in *HPV Nieuws* called for more attention to the theme of "HPVs and commuting." The author suggested the creation of an inventory of infrastructural bottlenecks for recumbent cyclists, so that the club, perhaps in cooperation with the national utility cyclists' organization, could address these issues with local authorities.[64] A few years later (in 1997 and again in 2000), similar arguments were articulated in the club's magazine, now connected to the idea of creating "super cycle paths," or bicycle highways, specifically designed for higher speed and long-distance bicycle commuting. In a 2000 article in *HPV Nieuws*, one member suggested that one reason why recumbents did not spread more among the population was exactly the lack of "a network of rapid, long distance cycle paths."[65] In the same year (significantly the year that the *Bike 2000 Construction Contest* ended in disillusionment), the NVHPV established an Infrastructural Commission to deal with issues like these. In 2006, at the occasion of local elections in the Netherlands, the most politically tinged issue ever of the club magazine was published, not however dealing with the politics of vehicle design, but being almost completely dominated by issues of cycling infrastructure.[66]

Not everyone welcomed the development. In 1995, NVHPV chairman (and *Fiets* editor) Guus van de Beek, who had been instrumental in organizing the 365-day bicycle contest, assessed the trend correctly as a step away from the original HPV ideas. He warned that the NVHPV should not become a second *Fietsersbond*—the national cyclists' interests organization—but instead should keep its focus on improving the "vehicles" themselves.[67] His warnings were of no use, however. The shift was not so much a matter of choice as the direct result of the majority of members becoming "consumer-riders" rather than "builder-riders," and of the influence of the dominant conception of bicycle promotion in the Netherlands, focused on infrastructural facilities for cyclists.

The Importance of Cultures of Cycling

The differences between the American and Dutch HPV clubs can be best understood in terms of their respective national cycling cultures.[68] In essence, whereas in the United States, the bicycle is considered marginal to or even opposing predominant aspects of American culture, in the Netherlands, cycling is commonly perceived as a part of Dutch national identity.

After the bicycle boom of the 1890s, for reasons not yet fully understood, the American market for adult bicycles declined sharply and eventually became insignificant until the early 1970s.[69] Typically, in the early 1950s, David Gordon Wilson (as he once noted) was considered a "very queer fish" when he, as an adult British immigrant, started to ride his bicycle in the States.[70] Around 1970, the bicycle was still essentially a children's vehicle: of the 7 million bicycles sold in that year, 5.5 million were for children.[71] After the 1970s boom, the bicycle also became a recreational and sports vehicle for adults. The percentage of people who participated in cycling for recreation more than tripled from 1960 to 1983, establishing touring bicycles as an important segment of the adult bicycle market at the end of this period.[72] But the bicycle was still hardly used for transportation, and in the perception of many Americans, utility cycling remained "an activity of odd people."[73] Although bicycle use increased somewhat in the last decades of the twentieth century, it never accounted for more than 1 percent of all trips.[74] Cycling infrastructure is not a standard ingredient of American road planning and construction, and a network of cycling infrastructure only exists in few US cities.[75]

A few examples may help to put flesh on the statistics. At the end of the 1990s, David Gordon Wilson was interviewed by a Boston newspaper about his passion for bicycle innovation. Two of the eleven (unprovoked) questions were: "So, do you hate cars?" and "Has your identification with cycling harmed your career?"[76] The lack of any explanation for these questions indicated the common understanding that there was somehow a negative correlation between bicycles on the one hand and cars and careers on the other.

Another example can help to clarify these associations. In the immensely popular Steven Spielberg film *E.T. the Extra Terrestrial* (1982), the bicycle plays a prominent role as the preferred vehicle of children, the heroes of this story. In the film's final chase scene, the children use their bicycles to successfully save the endearing E.T. from the hands of adult scientists and nasty government agents in cars, eventually launching into the air on their bicycles. In *E.T.*, the bicycle is therefore a slightly subversive vehicle tinged with romanticism, a "carrier" of the opposition against the grown-up world, authorities, and cars.[77]

As a subversive vehicle the bicycle was loved and respected in American counterculture—and equally distrusted by those who did not belong to this subculture. In the 1970s, it became a favorite argument of critics of the American way of life to point out that the motor-driven American army was defeated by the bicycle-riding Viet Cong, thus putting the bicycle in opposition to received ideas about American national identity.[78] The exhibition of a Ho Chi Minh Trail bicycle in the Smithsonian National Museum of American History arguably reinforces this "anti-American" image of the bicycle. Incomprehensible and incongruous to anyone unfamiliar with the specifics of American bicycle culture, the later HPV prominent Richard Ballantine dedicated his bestselling 1972 "manual of bicycle maintenance and enjoyment" to the anti–Vietnam War activist and bomb-setting terrorist "Samuel Joseph Melville, Hero."[79]

In short, for many Americans, cycling was (and is), if not a children's pastime or a sport, an expression of an alternative lifestyle, a consciously chosen deviation from the American identification with the motor car as the prototypical vehicle associated with the American Dream.[80]

For the Dutch, cycling had (and has) a completely different meaning, as comparative statistics about bicycle use in both countries may start to demonstrate. As in all European countries, the rise of automobility in the 1950s and 1960s had led in the Netherlands to a declining use of the bicycle, but compared to other nations, the Dutch still had relatively high levels of bicycle use when the bottom line was reached around 1970. By 1980, the bicycle's share in the modal split was about 27 percent, compared to fewer than 1 percent of all trips in the United States.[81] From the 1970s onward, the combined effort of new bicycle advocacy organizations and new government policies helped to reverse the downward trend and bicycle use began to increase again, both for daily use and (as elsewhere) for recreation.[82] By 2007, the use of the bicycle, measured in average distance traveled by bicycle per person each day, had increased from 1.5 kilometers in 1978 to 2.49 kilometers, whereas the same average for Americans in 2000 was still not higher than 0.1 kilometer a day.[83]

Contemporary observers describe Dutch cycling as a "national habit," instead of as a manifestation of a subcultural lifestyle.[84] Cycling is an activity performed on a daily basis by Dutch of all ages, occupations, political convictions, education levels, income groups, both sexes, and by both those who own cars and those who do not.[85] The omnipresence of bicycle paths and other infrastructural facilities for cyclists—on average, there is one kilometer of *fietspad* on every square kilometer of Dutch land—not only facilitates cycling and makes Dutch cyclists to a large extent feel safe from motorized traffic, but also effectively purveys the symbolic message to all road users that cycling is considered a public good.[86]

Cycling is considered common—except in the pejorative meaning of that term. On the contrary, the statement that *all* Dutch cycle is meant to convey that Dutch elites also do. Significant in this respect is the Dutch Royal family, who, since the 1930s, has upheld the tradition of showing themselves in public riding bicycles. The same is done by other representatives of Dutch elites, including ministers and prime ministers—both from the right and the left—who intend to show that they are not different from other Dutch citizens. It is perhaps remarkable that, given contemporary discussions, in the Netherlands compared with many other Western countries, cycling is not primarily associated with green or alternative lifestyles or specific political ideologies. Typically, although the present-day major Dutch bicycle advocacy organization, the Fietsersbond (est. 1975), started out as an organization with environmentalist goals, like many other cyclists' movements in the Western world that sprung up at the same time, it soon started to feel uncomfortable about these "ideological feathers" and in the 1990s removed references to environmentalist aims from its statutes.[87] In general, most of the time the Dutch do not consider cycling a conscious choice or a topic of reflection, but rather as a self-evident practice and a daily habit. Whereas in the United States, being "a cyclist" is often considered to be part of someone's identity, in the Netherlands, someone stops being a cyclist as soon as they get off their bicycle.

Contemporary Dutch bicycle culture is rooted in a long and persistent tradition.[88] As historical research has shown, in the interwar period, cycling became connected with Dutch national identity and acquired the status of a national symbol. That the bicycle has acquired this national status was at least in part due to the activities of the Dutch tourist and traffic organization ANWB, founded in 1883 as a cyclists' organization.[89] This influential lobby organization not only associated the bicycle consistently with love of the country—and therefore promoted from early on the construction of recreational bicycle paths—but also emphasized that cycling was a means to educate the masses and strengthen "national" character traits such as balance, discipline, and temperance among the population. At the same time, in 1905, the ANWB contributed to a law that forbade road racing. As a result, a strong (workers) culture of competitive cycling that thrived in some other European countries in the first half of the twentieth century did not develop here. Consequently, it could not challenge the middle-class values of quiet and respectable cycling.

When the bicycle spread under the working population and the motor car started to become available to the middle class members of the ANWB, the organization did not turn its back on the bicycle, but emphasized the different qualities of each vehicle type. Unlike the British Cyclists' Touring Club (CTC), for instance, which resisted the obligation to carry red rear lights and resented separate paths for cyclists, the ANWB did not take a polarizing stance against motorists. Even today, most Dutch do not consider riding bicycles and driving cars ideologically incompatible activities, but do both at different times, in different situations, and for different purposes.[90] The ANWB was keen to present itself as an organization of responsible citizens and negotiated pragmatically about measures affecting cyclists, for instance successfully demanding the construction of bicycle paths by the government in return for a new tax on bicycle ownership that was introduced in 1924 (although one can debate the fairness of this "deal," see the contribution by Albert de la Bruhèze and Oldenziel in this volume).

Dutch bicycle culture has materialized in a particular common Dutch bicycle design that is different from bicycle designs elsewhere.[91] Since the Dutch bicycle industry acquired a dominant position on the internal market during the interwar period, an indigenous type of bicycle came into being that is easily recognizable as such and popular until now (called in German the *Hollandrad*). The typically Dutch bicycle has a very upright, almost vertical, rider position, and, in accordance with normal Dutch usage, is fully equipped for all practical purposes—shopping, commuting, going to school, and visiting the city center—in all seasons. This bicycle design fitted the middle-class ideals of respectable and practical cycling perfectly. Except for developing this relatively slow and heavy machine for everyday use, the Dutch bicycle industry is not particularly known for its inventiveness. The Dutch industry was not involved in any of the major inventions related to the bicycle—draisine, vélocipède, safety bicycle, pneumatic tire, freewheel, gearing, recumbent bicycle, mountain bicycle, and e-bicycle. In this respect, Dutch bicycle firms have proven to be followers rather than innovators or trendsetters.

Relating Dutch bicycle culture back to the HPV movement and its American origins, one might suspect already that it was not a perfect match. True, in some respects the Netherlands *did* offer good circumstances for the HPV movement, providing an already cycling-prone population, a potentially huge market for new bicycle products, and a manifest willingness to invest in the promotion of cycling. However, although the Dutch HPV club in itself was not without success, as we have shown before, recumbents and other HPVs apparently had (and continue to have) only a limited appeal to the Dutch cycling public. A recent estimate suggests that there were in 2009 still only about 50,000 Dutch who own recumbent bicycles, tricycles, or "velomobiles," a number that stands in stark contrast to the more than 13 million bicycle owners in the country.[92] For Dutch HPV enthusiasts who were convinced of the technical superiority of their machines, the lack of public appeal was a recurrent source of frustration—and puzzlement. But on closer inspection it was hardly surprising.

To begin with, the Dutch bicycle sector is traditionally not geared toward innovation. Furthermore, the extraordinary bicycle designs of the HPV movement contrast with a national culture in which cycling is fully normalized and has even come to define normality. Even more than racing bicycles, the traditional upright Dutch bicycle design is almost the opposite of the low-slung horizontal recumbent designs of the HPV movement. Furthermore, Dutch cyclists do not associate cycling in the first place with sport or (lack of) speed. Neither do many Dutch feel that their regular bicycle is unsafe or uncomfortable. American concerns within the IHPVA over these issues seem to have taken the racing bicycle as a point of departure, with its uncomfortable rider position and its risk of being launched over the handlebars in case of heavy braking. Rather than seeing innovative HPVs as an answer to problems with safety, many Dutch consider recumbents and velomobiles themselves to be unsafe, both because of their higher speeds, and because their low profile gives the rider less overview and a higher risk of being overlooked by other traffic.[93] While it may be true that recumbents and velomobiles represent unbeatable racing machines, while they may provide unique solutions for cyclists with particular health problems (especially related to the wrist, back, and keeping balance), and although they can allow for longer distances (or shorter travel time) in commuting and may offer more comfort to long-distance tourists, for the multifaceted daily urban use of bicycles by most Dutch, recumbents or velomobiles may not be the most practical human-powered vehicles. Furthermore, whereas the IHPVA, from its American perspective, looked for a bicycle that could compete with or even replace cars, for the majority of Dutch people there was no urgent competition between the two, but rather a long-standing coexistence in which each vehicle type had proven to have particular benefits and setbacks. The anticar connotations of the term *human-powered vehicles* were contrasting with the habitual choice of most Dutch for the *fiets* as a practical and common mode of transportation not opposite, but next to, the car. Finally, whereas American HPV enthusiasts may have hoped to use *science* to elevate the bicycle from the limited spheres of childhood and alternative subcultures to more universal usage, from the perspective of almost universal Dutch bicycle use, there was no need for such an "elevation" in the first place.

Conclusion

The international history of the HPV movement presents an example of the diffusion and glocalization of ideas (or, rather, one idea) to promote cycling in the late twentieth century. This essay shows that such ideas may travel and spread easily, but are likely to change character and meaning in different cultural contexts that have evolved over longer periods of time. Anyone interested in promoting bicycle use should take these different cultural contexts of cycling into account and realize the different national trajectories through which they came into existence.

The context of Dutch national cycling culture not only limited the appeal of original HPV ideas in the Netherlands, but also changed the character and ideas of the Dutch HPV movement itself. The NVHPV developed from a club that, just like the IHPVA, was fascinated by the possibilities that technical innovation of the bicycle could offer, to a club that focused on the interests of recumbent riders, in particular by sharing experiences and product information and by demanding specific infrastructure. Over time, its membership changed from being mainly experimenters and builders into being predominantly users of commercial products. Most important of all, the Dutch HPV club was less evidently politically motivated than its American counterpart, and developed as much in opposition to "ordinary" bicycles as in opposition to the car. Typically, whereas the IHPVA explicitly embraced all human-powered vehicles and "all rider positions," within the NVHPV, more exclusively focused on recumbents, participation of a standard time-trial bicycle at HPV races could lead to heated discussions.[94] At the same time, within the NVHPV there seems to have been stronger skepticism concerning the idea of building a pedal-powered competitor to the car. As the prominent Dutch NVHPV member and velomobile-producer Ymte Sijbrandij once explained: "A velomobile is a very good bicycle, but a bad car."[95]

The latest indication for the different orientations of the American and Dutch HPV clubs is the growing unease within Dutch HPV circles as to the name of the club, which was originally suggested by an American-Dutch founding member. From the very beginning, the Dutch club was informally known as the *ligfietsvereniging*, or recumbent bicycle club, indicating that it was not so much preferring human power for propulsion that distinguished its members, but their preference for recumbent bicycles instead of ordinary bicycles.[96] Without strong proponents of the broader ideals connected with the concept, many members of the Dutch club consider the term *human-powered vehicles* too vague and not distinctive enough. It seems only a matter of time before this ongoing discussion will result in a change of the official name. Up to now, the debate has subsequently led to a change in the name of the club magazine from *HPV Nieuws* to *Ligfiets&* (the ampersand indicating that the magazine in principle might pay attention to other types of human-powered vehicles as well) in 2002, to the official recognition of Ligfietsvereniging NVHPV as an alternative name for the club in 2011, and to the adoption of a new logo, exclusively focused on recumbents, in 2012.[97]

From a comparative perspective, these changes indicate a step further away from the original HPV ideals as promoted by the IHPVA. The NVHPV, which owed its existence to the diffusion of the American, academic, and politically loaded concept of "human-powered vehicles," eventually had to adapt to the dominant Dutch cycling culture. From a hardcore HPV perspective this development may have been assessed as an essential loss. On the other hand, what is the loss if millions of Dutch ride HPVs every day, albeit of a less innovative sort?

As the history of the HPV movement shows, concerns over the environment have provided important incentives to the development of new bicycle designs since the 1970s and to the mobilization of expertise behind these products. The comparison of the HPV movements in the United States and the Netherlands demonstrates the importance of collective ideas and beliefs, not only in bringing about changes in mobility practices and contributing to the Western bicycle renaissance of the last few decades, but also in their powerful resistance to change. Convictions and beliefs, far from being "nonrational" hindrances to successful product development, in fact provided strong rationales for product development, as was shown to be the case for both the recumbent bicycle and *Hollandrad*.[98] Paradoxically, that being said, in the context of Dutch bicycle culture, a nonideological, pragmatic approach to bicycle innovation may indeed lead to the best responses from the public. Not because such an approach is "rational," but because it is in line with dominant characteristics of Dutch bicycle culture.

Manuel Stoffers is assistant professor of history at the Faculty of Arts and Social Sciences of Maastricht University. His research interests include the history of cycling cultures and the history of cycling as a mode of mass transportation. He has published on cycling history in *Transfers, Mobility in History, Cycle History, Journal of Transport History* and in *BMGN/The Low Countries Historical Review*. In 2009, he launched an online international cycling history bibliography (www.fasos-research.nl/sts/cyclinghistory).

Notes

1. Stuart S. Wilson, "Bicycle technology," *Scientific American*, March 1973, 81–91, here 91.
2. See on the new bicycle activism in general: Benoît Lambert, *Cyclopolis, ville nouvelle: Contribution à l'histoire de l'écologie politique* (Genève, 2004); Paul Rosen, "Up the Vélorution: Appropriating the Bicycle and the Politics of Technology," in *Appropriating*

Technology: Vernacular Science and Social Power, ed. Ron Eglash et al. (Minneapolis, MN, 2004), 365–389; Dave Horton, "Environmentalism and the bicycle," *Environmental Politics* 15, no. 1 (2006): 41–58; Zack Furness, *One Less Car: Bicycling and the Politics of Automobility* (Philadelphia, PA, 2010). The history of Critical Mass is documented in Chris Carlsson, *Critical Mass. Bicycling's Defiant Celebration* (Oakland, CA, and Edinburgh, 2002).

3. For non-Western perspectives, see the "Special section on Global Cycling" of *Transfers* 2, no. 2 (2012): 22–126, edited by Ruth Oldenziel and Adri A. de la Bruhèze, containing articles on Japan, China, West Africa, and Finland.

4. Quote from J. Harry Wray, *Pedal Power: The Quiet Rise of the Bicycle in American Public Life* (Boulder, CO, and London, 2008), 168–169, see also 11. The first time US federal funds were made available for bicycle projects was in 1991 when the US Congress passed the ISTEA legislation; in 2008, Congress passed a Bicycle Commuter Benefits Act.

5. For a general historical assessment of the HPV movement, see also Manuel Stoffers, "The Human Powered Vehicle Movement and the Changing Image of the Bicycle at the End of the Twentieth Century," in *Cycle History: Proceedings of the International Cycling History Conferences* 22 (2012): 211–219. Older histories of the IHPVA are: [Anonymous], "The History of IHPVA," *Human Power* 1, no. 1 (1977): 10, 19; Chester R. Kyle, "A Brief History of the International Human-Powered Vehicle Association 1976–1998," *Cycle History: Proceedings of the International Cycling History Conferences* 12 (2001): 134–145; Kyle, "A history of human-powered land vehicles and competition," in *Human-Powered Vehicles*, ed. Allan V. Abbott and David G. Wilson (Champaign, IL, 1995), 95–111; Arnfried Schmitz and Tony Hadland, *Human Power: The Forgotten Energy 1913–1992* (Coventry and Le Thor, 2000).

6. Chester Kyle, "Bicycle Aerodynamics and the Union Cycliste Internationale," *Cycle History: Proceedings of the International Cycling History Conferences* 11 (2000): 118–131; Rosen, "Up the Vélorution"; Arthur Eger et al., "Mislukt: Roulandt ligfiets," in *Productontwerpen*, ed. Eger et al. (Utrecht, 2004), 237–239; Luca Berchicci, *The Green Entrepreneur's Challenge: The Influence of Environmental Ambition in New Product Development* (PhD thesis, Delft University of Technology, 2005); Peter Cox, "The Role of Human Powered Vehicles in Sustainable Mobility," *Built Environment*, 34(2) (2008): 140–160.

7. A. C. Gross, Chester R. Kyle, and D. J. Malewicki, "Aerodynamics of Human-Powered Land Vehicles," *Scientific American* December (1983): 126–134. The article by Stuart S. Wilson on "Bicycle Technology" in *Scientific American* of 1973 had not indicated any need for technical improvements (although the author did mention efforts to create a human-powered aircraft).

8. Compare Rosen, "Up the Vélorution," 377.

9. See Andrew Kirk, "'Machines of Loving Grace': Alternative Technology, Environment, and the Counterculture," in *Image Nation: The American Counterculture of the 1960s and '70s*, ed. Peter Braunstein and Michael W. Doyle (New York and London, 2002), 353–378, here 367.

10. For figures on the bicycle renaissance, see John Pucher, "Bicycling Boom in Germany: A Revival Engineered by Public Policy," *Transportation Quarterly: An Independent Journal for Better Transportation* 51 (1997): 31–46; John Pucher and Ralph Buehler,

"Making Cycling Irresistible: Lessons from the Netherlands, Denmark and Germany," *Transport Reviews* 28, no. 4 (2008): 495–528. See for a discussion on the bicycle renaissance Manuel Stoffers and Anne-Katrin Ebert, "New Directions in Cycling Research: A Report on the Cycling History Roundtable at T2M Madrid," in *Mobility in History* 5, ed. Peter Norton et al. (New York and Oxford, 2014), 9–19, here 13–16.

11. Frank J. Berto, "The Great American Bicycle Boom," *Cycle History: Proceedings of the International Cycling History Conferences* 10 (1999): 133–141.
12. E.g., A. J. Hand, "Pedal-Powered Cars ... Is That a Way to Go?" *Popular Science*, August 1973, 100–101; Sheldon M. Gallager, "Pedal Cars: The Gasless Way to Go," *Popular Mechanics*, May 1974, 98–99.
13. Among the popular and scholarly publications on the history of the mountain bicycle, the atmosphere in which it was invented is still best captured in the eyewitness account by Charles Kelly (and Nick Crane), *Richard's Mountain Bike Book* (Sparkford, 1988), esp. 19–55.
14. Kyle, "A Brief History," 137–138.
15. Kyle, "A history of human-powered land vehicles and competition," 103–104.
16. In 1934, the Union Cycliste Internationale (UCI), the world governing body of bicycle racing, disqualified an hour record made on a recumbent bicycle and defined strict rules for the obligatory geometry of bicycles participating in UCI races. In the last decades of the twentieth century, the UCI has again restricted technical innovation of bicycles; see Kyle, "Bicycle Aerodynamics and the Union Cycliste Internationale," 127.
17. [Anonymous], "The History of IHPVA," 10.
18. Chester Kyle, "Supercycles," *Omni*, July 1979, 97–100, 120, here 97.
19. See David G. Wilson, "A plan to encourage design improvements in man-powered transportation," *Engineering* 204, no. 5283 (21 July 1967): 97–98.
20. Frank R. Whitt and David G. Wilson, *Bicycling Science: Ergonomics and Mechanics* (Cambridge, MA, 1974), esp. 205–233.
21. E.g., David G. Wilson, "The future of the bicycle and its potentialities," *Long Range Planning* 8, no. 2 (April 1975): 81–87, here 84; David G. Wilson, "The future potential for muscle power," in *Pedal Power in Work, Leisure, and Transportation*, ed. James C. McCullagh (Emmaus, PA, 1977), 106–124, here 108–110. On the pre- and postwar history of the recumbent bicycle, see also Gunnar Fehlau, *Das Liegerad* (Kiel, 1996), esp. 9–29 (published in English as *The Recumbent Bicycle*), and Tony Hadland and Hans-Erhard Lessing, *Bicycle Design: An Illustrated History* (Cambridge, MA, and London, 2014), 473–492.
22. The final prototype Avatar 2000 was finished and patented in 1979 and shown in Europe in 1980; see David G. Wilson, "The development of modern recumbent bicycles," in *Human-Powered Vehicles*, ed. Abbott and Wilson, 113–127. The first Avatars apparently were produced early March 1981; see R. K., "The bike that sells itself," *East West Journal* (11 May 1981): 28.
23. David G. Wilson, "A blue print for an HPV revolution," in *Third International Human-Powered Vehicle Scientific Symposium. August 28 and 29, 1986. Proceedings*, ed. Allan V. Abbott (Indianapolis, IN, 1986), 79–83, here 79.
24. See David G. Wilson, "Man-powered land transport," *Engineering* 207, no. 5372 (11 April 1969): 567–573, here 567.
25. Whitt and Wilson, *Bicycling Science*, 232.

26. Richard Ballantine, "The politics of human-powered vehicles," *Human Power* 54 (2003): 3. American Ballantine was a founding member of the British HPV club and author of *Richard's Bicycle Book*, which, since its first edition in 1972, sold more than 1 million copies, was translated into German, French, Spanish and Dutch, and included information on bicycle activism, bicycle innovation, and HPVs.
27. Wilson, "Blue print for an HPV revolution," 79; see also Wilson, "The future of the bicycle."
28. Kyle, "Supercycles," 120.
29. See *Human Power* 8 (Spring 1982): 2; see also Allan Abbott, "Practical vehicles at last!" *Human Power* (Spring 1983): 8.
30. Kyle, "A Brief History," 137.
31. Ibid., 143.
32. Abbott and Wilson, *Human-Powered Vehicles*.
33. In its first issue, *Human Power* 1 (Winter 1977/78): 6–7 and 19, could list over sixty articles on the HP speed championships in forty different periodicals and six languages, including German and Russian.
34. I am not aware of any direct connections with Stewart Brand, but David Wilson did contribute to *Pedal Power*, ed. McCullagh, which was a typical publication of the appropriate technology school (to which Stuart S. Wilson, author of the 1973 *Scientific American* article on bicycle technology, also contributed). Also, in the IHPVA journal *Human Power*, several contributions were published on HPVs in developing countries. See on Brand and the AT-movement, Kirk, "'Machines of Loving Grace.'"
35. The 1980 statutes of the IHPVA said that its specific purpose was to "encourage and support unrestrictive, innovative and creative design and construction of vehicles solely operated by human power for transportation on land, in the air and on the water"; see *Articles of Incorporation of International Human-Powered Vehicle Association* (1980), www.ihpva.org.
36. Bruce D. Epperson, "Bicycle Planning, American Style: A History of Vehicular Cycling, 1968–1982" (version 4, 2012): 42, www.bikereconstruction.com/Documents/History_of_Vehicular_Cycling_4.pdf (accessed 17 September 2014).
37. See James Longhurst, "The sidepath not taken: Bicycles, taxes, and the rhetoric of the public good in the 1890s," *The Journal of Policy History* 25, no. 4 (2013): 557–586; Epperson, "Bicycle Planning, American Style."
38. John Forester, *Bicycle Transportation* (Cambridge, 1983), 3.
39. Kyle, "Supercycles," 97 (italics in original).
40. Initially, there were some complaints about the Californian-centeredness of the IHPVA; see e.g., *Human Power* 8, vol. 2, no. 1 (1982): 4. Later there were recurring complaints that the IHPVA was too American and not sufficiently international; see e.g., *Human Power*, 9, nos. 3–4 (Fall/Winter 1991/2): 2. For this reason, in 1998 the IHPVA was renamed HPVA and a new IHPVA was established, an umbrella organization formed out of national HPV clubs; see *Human Power* 52 (Summer 2001): 23. After continuing conflicts, the American-based HPVA in 2008 reclaimed its original name, and the IHPVA was renamed W(orld)HPVA. See www.whpva.org.
41. See Kyle, "A Brief History," 137.
42. At the latest since the edition of 1984, Richard Ballantine integrated information about recumbent bicycles and other innovative HPVs in his *Bicycle Book*.

43. See *Human Power* 1, no. 1 (1977): 6, 19. In the Lithuanian SSR, engineer and university professor Vytas Dovydėnas was a leading figure from the 1970s onward; see his book *Velomobile* (Berlin, 1990), translated from the Russian version that appeared in Leningrad in 1986; earlier, Wilson's *Engineering* contest from 1967 had already invited several Polish entries. In 1987, a Polish HPV club called the Circle of Alternative Vehicles was established as a section of the Polski Klub Ekologiczny; see the letters by Marek Utkin in *BHPC Newsletter,* February 1987, 6, and July 1987, 23.
44. See *Human Power* 2 (Winter 1978): 7; see Schmitz and Hadland, *Human Power,* 43ff, for an eyewitness account of the atmosphere at the early HPV races in Europe. See Dovydėnas, *Velomobile*: 7, 19–20, 25–33, on the Šiauliai design competition and references to American developments.
45. On the Dutch doubts about the feasibility of a magazine devoted to cycling, see Guus van de Beek, "5 jaar," *Fiets* 5, no. 2 (March 1987): 7.
46. See, e.g., the report on British HPV races in *Fiets* 1, no. 5 (1982): 13–17; a report on the Zandvoort races appeared in *Fiets* 2, no. 6 (Nov/Dec 1983): 20–26; on Peterson, see *Ligfietse&* 21, no. 2 (2005): 6–7. Already in 1980, *Pedal Power,* ed. McCullagh, containing a contribution on recumbents by David Gordon Wilson, had been translated in Dutch and was published as *Trapperkracht.* Richard Ballantine's *Bicycle Book* was translated as *Het groot Fietsenboek* and published in 1992.
47. See "Verslag van de oprichtingsbijeenkomst van de Nederlandse Human Powered Vehicle Vereniging op 29-9-1984 te Aalsmeer," http://oud.ligfiets.net/nvhpv/notulen.php3?id=67.
48. "Mike Burrows reports from Britain," *Human Power* 15, vol. 5, no. 1 (1985): 6.
49. The following is partly based on Manuel Stoffers, "Historische achtergrond van de NVHPV," *Ligfietse&* 26, no. 5 (2010): 10–15.
50. See http://oud.ligfiets.net/nvhpv/statuten.php3; compare footnote 35 for the IHPVA statutes.
51. See *Ligfietse&* 26, no. 5 (2010): 9. Membership after 2002 declined somewhat and was about 1710 in 2010.
52. See on the market introduction of the Roulandt: Eger, "Mislukt: Roulandt ligfiets." See on other Dutch recumbent firms the series of interviews by Henk Zwols with Dutch recumbent producers in *HPV Nieuws* (1999–2002); for data on Velomobiel, see Berchicci, *The Green Entrepreneur's Challenge,* 133–149. Many of the early Dutch HPV producers later explained that they were first introduced to the HPV phenomenon through 1982–1983 articles in *Fiets* and *Scientific American*; this is documented, e.g., for Bram Moens (M5), Johan Vrielink (Flevobike), Jan Eggens (JE), Jouta, Valenteyn.
53. The Dutch bicycle firms Batavus (model Relaxx, 1997) and Sparta (models Icarus and Piraeus, 1999) experimented with recumbents on the basis of models developed by recumbent firms; bicycle firm Gazelle developed the Easy Glider 'zitfiets' as part of the Mitka project in 2004; see Berchicci, *Green Entrepreneur's Challenge,* 125–128.
54. *HPV Nieuws* 5, no. 2 (1989): 25.
55. "Notulen van de Algemene Ledenvergadering NVHPV 2001," oud.ligfiets.net/nvhpv/notulen.php3?id=21
56. See Ton Welleman, *Eindrapport Masterplan Fiets. Samenvatting, evaluatie en overzicht van de projecten in het kader van het Masterplan Fiets, 1990–1997* (Den Haag, 1998), 96.

57. Berchicci, *The Green Enterpreneur's Challenge*, esp. 100–132.
58. *HPV Nieuws* 1, no.1 (1984): 1, 2.
59. *HPV Nieuws* 2, no. 4 (1985): 4.
60. *HPV Nieuws* 5, no. 2 (1989): 14–23.
61. On the symposium, see the report in *HPV Nieuws* 9, no. 3 (1993): 6-9. In 1985, Van Ingen Schenau had co-edited a Dutch textbook on scientific approaches to competitive cycling, in which he also referred to recumbents and streamlined HPVs; see *Wielrennen. Een confrontatie tussen de wetenschap en de praktijk van het wielrennen*, ed. Jan-Pieter Clarus and Gerrit Jan Van Ingen Schenau (Lochem and Gent, 1985), esp. 232–236.
62. *HPV Nieuws* 11, no. 1 (1995): 13.
63. *HPV Nieuws* 15, no. 5 (1999): 18–19.
64. *HPV Nieuws* 9, no. 6 (1993): 23.
65. *HPV Nieuws* 16, no. 5 (2000): 2–3.
66. *Ligfiets&* 22, no. 1 (2006).
67. *HPV Nieuws* 11, no. 1 (1995): 4.
68. Cf. Peter Pelzer, "Bicycling as a Way of Life: A comparative case study of Bicycle Culture in Portland and Amsterdam" (MA thesis in metropolitan studies, University of Amsterdam, 2010); For the United States, see also Furness, *One Less Car*, and Wray, *Pedal Power*.
69. David V. Herlihy, *Bicycle: The History* (New Haven, CT, and London, 2004), 309–342.
70. David G. Wilson, "Getting in gear: Human-powered transportation," *Technology Review* 81 (October 1979): 42–54, here 42.
71. Frank J. Berto, *The Dancing Chain: History and Development of the Derailleur Bicycle* (San Francisco, CA, 2010), 230.
72. H. Ken Cordell, "Outdoor recreation participation trends," in *Outdoor Recreation in American Life: A National Assessment of Demand and Supply Trends*, Cordell et al. (Champaign, IL, 1999), 219–321, here 235; Berto, *The Dancing Chain*, 245–246.
73. John Forester, "The place of bicycle transportation in modern industrialized societies" (paper presented to the Preserving the American Dream Conference, 2005), www.johnforester.com/Articles/Social/place_of_bicycle_transportation.htm.
74. See John Pucher, Charles Komanoff, and Paul Schimek, "Bicycling Renaissance in North America? Recent Trends and Alternative Policies to Promote Bicycling," *Transportation Research Part A* 33, no. 7/8 (1999): 625–54, here table 1.
75. Ralph Buehler, John Pucher, and Uwe Kunert, "Making Transportation Sustainable: Insights from Germany. Prepared for the Brookings Institution Metropolitan Policy Program" (2009): 21.
76. John Koch, "David Gordon Wilson," *The Boston Globe*, 7 June 1998.
77. Interestingly, the scene with the flying bicycles echoes the many late-nineteenth-century advertising posters that depict cyclists effortlessly flying through the air.
78. The *locus classicus* for this statement is Ivan Illich's *Energy and Equity* (London, 1974).
79. Richard Ballantine, *Richard's Bicycle Book* (New York, 1972), [II]. Melville was found guilty of having organized eight bombings of major private and public institutions in New York in 1969; he was killed in 1971 during a prisoners' revolt that he had helped to organize.
80. See Wray, *Pedal Power*, esp. chap. 3: "Culture Storm."

81. See Pucher, Komanoff, and Schimek, "Bicycling Renaissance in North America?" table 1; Welleman, *Eindrapport Masterplan Fiets*, 44.
82. Welleman, *Eindrapport Masterplan Fiets*; Pucher and Buehler, "Making Cycling Irresistible."
83. Statline.cbs.nl ("Mobiliteit per regio, vervoerwijzen, persoonskenmerken, 1985–2007"); Pucher and Buehler, "Making Cycling Irresistible," 497 and 503.
84. Giselinde Kuipers, "The Rise and Decline of National Habitus: Dutch Cycling Culture and the Shaping of National Similarity," *European Journal of Social Theory* 16, no 1 (2012): 17–35; Pelzer, "Bicycling as a Way of Life."
85. Exact statistics are available on Statline.cbs.nl ("mobiliteit per regio, vervoerswijzen, persoonskenmerken, 2007"); see also Pucher and Buehler, "Making Cycling Irresistible," 502–505.
86. Best statistics on the total length of Dutch bicycle paths are made available by the Fietsersbond, see www.fietsersbond.nl/nieuws/bijna-35000-km-fietspad-nederland#.VBmFJvl_tBk (accessed 17 September 2014).
87. Bram Duizer, *"In het nut van actie moet je geloven": dertig jaar actievoeren door de Fietsersbond* (Utrecht, 2005).
88. Manuel Stoffers, "Cycling as Heritage: Representing the History of Cycling in the Netherlands," *Journal of Transport History* 33, no. 1 (2012): 92–114.
89. See for this and the following paragraph: Anne-Katrin Ebert, *Radelnde Nationen. Die Geschichte des Fahrrads in Deutschland und den Niederlanden bis 1940* (Frankfurt, 2010).
90. See Statline.cbs.nl ("Mobiliteit per regio, vervoerwijzen, persoonskenmerken, 1985–2007"). In 2007, on average the Dutch (including university and high school students and pupils) rode 2.49 km/day; for Dutch car owners the distance ridden by bicycle was on average still 1.81 km/day (while those without car or driving license rode on average 2.88 km/day).
91. See Anne-Katrin Ebert, "Nationales Design? Auf der Suche nach dem "Holland-Rad," 1900–1940," *Technikgeschichte* 67, no. 3 (2009): 211–231; Herbert Kuner, "De balhoofdbuislengte van een omafiets," *De oude fiets* 4 (2009): 18–22. This fact is overlooked by Zahid Sardar, *The Dutch Bike. De Nederlandse fiets* (Rotterdam, 2012), a design history unfortunately fraught with mistakes.
92. See *Ligfiets&* no. 5 (2010): 16, and Statline.cbs.nl ("Voertuigbezit 2007").
93. There are few trustworthy surveys on the reception of recumbents; see Manuel Stoffers, "Ligfiets onveilig en onhandig. Delftse studenten onderzoeken meningen bij gebruikers en niet-gebruikers," *Ligfiets&* 22, no. 4 (2006): 34–35.
94. See *HPV Nieuws* 2, no. 3 (1985): 5; *HPV News* [IHPVA] 8, no. 1 (1991): 4.
95. Quoted by Berchicci, *Green Entrepreneur's Challenge*, 133.
96. See *HPV Nieuws* 2, no. 4 (1985): 5.
97. Embedded in a national cycling culture not unlike the Dutch, the somewhat younger HPV-Klub Danmark (est. 1993) changed its name already in 2007 to Liggecykelforeningen ("Recumbent Bicycle Association").
98. In a dissertation in which the commercial and public development of HPVs in the Netherlands is contrasted, Luca Berchicci has argued that "high level environmental ambition" tends to be detrimental to successful product development. See Berchicci, *Green Entrepreneur's Challenge*, 182.

 CHAPTER 3

Scarcity, Poverty, Exclusion
Negative Associations of the Bicycle's Uses and Cultural History in France

Catherine Bertho Lavenir

Today's urban policies predict a revival of the bicycle for commuting as well as for short- or medium-distance trips in central and southern Europe, and notably in France.[1] The return to the bicycle as mode of transport in France signals a turning point in a century-long evolution of mobility culture. This use of the bike had been all but discarded in French towns from the 1960s onward, to be replaced by motorized vehicles: the car, the motorbike, and the moped.[2] Thus the return to urban use will not necessarily be a smooth process. The typical criteria considered by riders when choosing a form of transport are very significant, yet they seem to have been underestimated by researchers. This chapter is an analysis of the negative associations within French society of the specific uses of the bicycle. I will also reflect on the role of memory in the social advent of mobility.

Symbolic Elements

When analyzing why users are motivated to access new forms of mobility, today's researchers tend to lean toward the objective considerations of the decision such as cost or efficiency; they focus on the practical reasons for adopting new, green solutions to traffic problems, like resorting to the bike in urban areas. Such an approach overlooks the subjective considerations, namely the representations bike riders have of themselves, consciously or not. I will examine the negative images and associations that either encourage users to discard a mode of transport, or make them reluctant to readopt it. These representations are instilled with individual and collective memory. Here, then, is my question: how has the collective memory of the bicycle's uses functioned in France in the second half of the twentieth century? I will examine in more detail one type of perception: the negative recollections linking this mode of transport with difficult times. The reason for this choice: in France there is a kind of endearing myth surrounding the bicycle, kept alive by the media, associating it with a supposedly golden age of popular and working-class culture, glorified in film and song.[3] But the social uses of the bicycle throughout the twentieth century reveal that this mode of transport was frequently associated with situations of want, scarcity, and danger. How did the collective memory engineer the recollection of such episodes, and to what extent could they obstruct the spread of new types of mobility? How were individual memories handed down within families? What was the role of the mass media in fixing images and events that have become common knowledge? Are we not in fact being confronted with a sort of reversed version of events, associating the genuinely painful memories of occupied France with positive recollections, while the not-so-difficult years of the 1950s and 1960s have endurably linked the bicycle with images of social misfits and poverty? To understand this point, I propose to closely examine the mirror play between reality and representation in the general setting of occupied France during World War II, a setting that cannot be properly evaluated if we ignore propaganda and its mendacious representations.

What follows is an investigation not only of the creation of memory around an object—the bicycle—but also of the mobility situations leading to its use in twentieth-century France.[4] Close examination of two historical periods lets us see how a complex relationship can arise between representation, transmission, and the elaboration of common memory impinging on today's mobility options.

The periods under review represent times when the bicycle occupied a privileged place in the country's everyday practices and economy. In twentieth-century France, two periods have become associated with negative perceptions of the bicycle. The first is the time of the German occupation from 1940 to 1945. The bicycle was then associated with situations of danger (avoiding air raids, "resistance") and scarcity (the quest for food). At the same time, these negative memories of the "occupation" come to mind in a positive context: that of communities and networks based on helping one another, with the bike as a tool serving solidarity and fellowship among families, friends, or larger groups. For various reasons, this line of collective memory does not seem to form an obstacle to the new urban mobility.

The second period I have examined is that of the 1960s and 1970s, when economic expansion meant the disappearance of the traditional peasant world, intense urbanization, and the potential for the lower classes to own cars. The people who continued to use their bikes for everyday transport were gradually cast to the margins of the dominant social norm. The urban use of the bike acquired permanently negative perceptions, notably in those "life stories" that tell the author's tale as one of upward social mobility. Being "condemned" to the bicycle became a stigma signaling failure in a social trajectory, an indication that the individual concerned had been left behind by his more affluent contemporaries.

In this chapter, I would like to develop the following hypothesis: In France, the bicycle suffered from negative representations at least twice during the second half of the twentieth century. Such negative images lasted long enough in the collective memory to constitute a specific obstacle to a general comeback of the bike in an urban context, at least for older generations. This might explain why a country like France was slow in adopting the same attitude to the bike as commonly held in northern Europe.

I will explore the aspects of the two periods proposed for analysis by first discussing the actual uses of the bicycle, and then by looking at the modes of representation prevalent in each period. Finally, I will discuss the trajectories and underpinnings of memory, thus reflecting on how a collective memory is formed.[5] I relied on three types of sources: interviews with people who had experienced the hardships of the German occupation as children or teenagers; autobiographical accounts or memoirs published by nonprofessional authors in traditional format or on the Internet; and films that have become part of French popular culture. My hypothesis is that such works, in this case films, reinforce how the perception of memories is formed.

The Bicycle in France under German Occupation

During the German occupation, many French people were compelled to undergo a form of technological regression. The use of the car was restricted to the occupiers and a few designated professions such as doctors or the police, who were the only people permitted to buy petrol. Besides, public transport was poor, and in 1944, disorganized as well as dangerous. For all these reasons, people who had abandoned the habit of using a bike in town had to return to this form of transport.[6]

Until 1939, the car had rarely been used to drive from home to work. In contrast, middle-class families had begun to use automobiles regularly for weekend outings to the country or to visit relatives.[7] Often those cars had been bought recently, and were seen as a symbol of social status. Now the car was being replaced by an assortment of family bikes, amateurishly or deftly adapted in order to carry young children or household pets as well as bags, baskets, and umbrellas.

Consider, for example, the comments of A. L. in an interview which reflect this wartime practice: "My father had a Rosengart. At the beginning of the war, he had left it in the garage with its wheels hidden away, so that it would not be confiscated. On Sundays we would go for a ride. My father had fixed a small saddle on the cross bar for me to sit on."[8]

Due to the same circumstances, white-collar employees who were used to going to work by tram, bus, or train had to resort to the bicycle when confronted with the chaotic public transport, thus reducing them back or down to the status of lower-class workers riding their bikes to the factory.[9]

In French cities and larger towns, and especially in Paris, another mode of transport emerged, based on the bicycle, whereby a man on a bicycle replaced the horse. Such vehicles were never used in large numbers, yet their very existence was symbolically significant. "Taxi-bikes" were comparable in principle to the rickshaw found in the colonies, where a native would transport a white man. The feeling of degradation experienced by these men is not something we should overlook. In France, like in other occupied countries, the journalists of the "collaborationist" newspapers presented the taxi-bikes in a picturesque vein, featuring humorous anecdotes, which were nevertheless saturated with ideological representations.[10]

Let us consider what the magazine *L'Illustration* wrote in 1941 in a special issue on the automobile. *L'Illustration* was a high-end publication featuring photographs, drawings, and watercolors, and it had definite right-wing tendencies. Since 1920, it had devoted an annual issue to cars, usually coinciding with the "Salon" exhibition. By 1941, the magazine had become one of the pillars of upper-class "collaboration."[11] An article by René Baschet, "Ré-inventer," which discusses the new mode of transport in Paris, took on board all the sexist, racist, and colonialist clichés of the time.[12] The subtext included all the themes of the collaborationist press.[13] It integrated and applied the stereotypes, depicting a nation vanquished but happy, its resourceful people eager to invent individual and limited responses to defeat and oppression.

The report on the taxi-bikes located in the Bois de Boulogne in Paris, the so-called *taxis 41*, uses humor to disguise a point of view that is both sexist and class-conscious. The bicycle is described as a "steel steed."[14] Pleasantly depicted at first, the cyclist riding the steel steed now assumes a woman's shape, inspiring an overtly sexist appreciation:

> Customers can, if they like, choose a single seat or tandem, an athletic or sturdy girl.... For indeed strong calves enable energetic maidens to ferry to and from Longchamp and Porte Maillot full cargoes of passengers ... gentlemen for the most part. It seems, indeed, that snug in his small carriage, a cigarette dangling from his fingers, a scarf around his neck, the male passenger takes particular delight in entrusting himself to the girl, for the sheer sight of a shapely back and nervous legs, or to enjoy some sort of subtle revenge. ... The cyclist towing their passenger is a "bizarre centaur" half animal: when

upright on their pedals, utterly disheveled, the strange centaurs are seen straining up a slope.

And yet, the feeling of individual and national humiliation creeps onto the page. When an arm thrusts out an umbrella to hail a taxi-bike: "This means that a humiliated passenger is in a hurry to get to the end of the route and deliver his driver to the group of knitting girls, who, wearing shorts and caps, congregate on benches while waiting for customers."

In this context, the cyclist is also a machine and the following image betrays a racist mindset: "Among them is a superb couple on a tandem: the male is a superb black man, the woman is white; and their long naked legs, so dissimilar, but both making exactly the same movement, make one think of a machine that someone has not finished painting."

A journalist could not ignore the colonial reference. The streets in Paris let him witness a "picturesque animation resembling certain market corners under exotic skies, where rickshaws collide, and men in gaudy clothing burst into insults."[15] A France humiliated, turned into a German colony, whose men were seen as animals or machines, and whose women were defenseless prey to the sexist eye: all of this could be found in the description of the taxi-bike, brimming with falsely genial sympathy.

According to another article in the same issue, the return of the bicycle enabled people to associate its use with the reactionary values of the "French State" (the collaborationist regime having replaced the Republic).[16] At first, the forced use of the bicycle was presented without the slightest trace of negative criticism, delicately skipping over its war origin. We can even find a discreet approval of the social leveling brought about by cycling, which might bring together a national community hitherto far too fragmented.[17]

If the bicycle was a means to get to work, it was also a way to find new and better forms of holiday-making better. It enabled the French tourist to return to the true values of the cycling trip, discovering the beauty of France; in short, to adopt the antimodernist ideology of the French State. The magazine article opened with a theme characteristic of collaborationist propaganda: a call for self-examination. The intention was to direct French people to accept the notion that they were responsible for the current national misfortune; this notion was reinforced by disparaging a string of modern attitudes, representations, and values identified as belonging to 1930s France, the love of speed being one of them.[18]

Praising slowness, presented as desired rather than forced upon people, went hand in hand with another cherished theme of Vichy propaganda: the necessity to love France and the need to know and discover the country in order to love it. As traveling abroad was forbidden, the right attitude, the magazine said, was to make the best of the situation and take to the roads of the homeland.[19] So all was for the better, except that a bizarre comparison with black slaves in Louisiana suggested there was more in the writer's mind than a mere report of a patriotic outing to the countryside.

The photographs published during the occupation in the collaborationist magazines—the only authorized ones—developed a visual comment about the bicycle. Let us consider the pictures taken by André Zucca.[20] Color films were reserved for the photographers authorized by the German occupiers, and Zucca's photographs were destined for the propaganda papers. There are numerous pictures of bicycles, taxi-bikes, bicycles with trailers, and all the paraphernalia connected with this French makeshift mode of transport.

The photos for *Signal*, a German propaganda magazine, showed the bicycles in occupied Paris in the same light as described above in the two *L'Illustration* articles. You notice the picturesque elements of Paris in the sun, where anything out of the ordinary—the presence of Nazi flags, the absence of cars—was seen as "incidental." Another picture by Zucca shows a group of young people going for a ride. The girls' white ankle socks suggested outdoor simplicity, not a shortage of stockings. All you can see on these enthusiastic cyclists' faces is the jolly comradeship of sportsmen, with no hint whatsoever at the reason why the largest square in the capital city was so strangely deserted.

Both German and Vichy propaganda offered ideologically loaded representations of the bicycle's uses. The actual experience of the French was definitely different to the one observed through a nationalistic, chauvinistic looking glass. Yet we seem to discern a kind of inversion through memory, which can be observed at different levels: in the works of fiction that create a common memory of events and within individual memoirs and autobiographical reports.

Happy Memories of Dark Times?

The uses of the bicycle during the war and occupation in France were connected with sad, sometimes tragic events, and were certainly surrounded by a difficult context. Yet, when screened through memory, the most terrifying and dangerous circumstances associated with the bicycle seemed to acquire positive connotations. Moments when a bicycle was part of the scene in a particular context of war came to be integrated in the collective memory through stories filtering their significance. The reevaluation of the recollected uses of the bicycle is inscribed within a larger transformation, resulting in the construction of a collective memory of "the occupation."

A sort of great national narrative took shape, allowing for a united perception of France under "the occupation," while the nation, in truth, was prey to deep divisions at the time. It must be remembered that during the occupation, the divide between "collaborationists" and "resisters" was a violent one; tens of thousands of members of the Resistance were deported, tortured, and shot. The end of the war saw large-scale purges (also called "purification") carried out via death sentences and executions. As is the case in any country undergoing such a symbolic partition, works of fiction were created for a long time afterward that were intended to rebuild national identity. Such fictitious reports are naturally euphemistic and present recollections linked to bicycles during the occupation in a different way.

One particular film, *La Grande Vadrouille* (1966), is emblematic of this later inversion of values. It tells of the forced association between two representative characters in French society, who are opposites in every way: a housepainter (played by the popular actor Bourvil) and a musical conductor (played by the equally popular Louis de Funès).[21] The two have somewhat reluctantly joined the Resistance to save British airmen. They are discovered and hunted down, and in order to escape, try to reach a safe place in southern France just beyond the demarcation line. The two actors were favorites with French audiences; here they offer a burlesque replay of a classic Resistance operation as a metaphor of French unity in the face of adversity. The story is reminiscent of Jean Renoir's successful *La Grande Illusion* (1937), whose prewar theme was exactly the same. We should note that the earlier film was made with the deliberate political aim of fostering a distinct French identity.

La Grande Vadrouille is considered one of the films produced around the 1960s that revisited the Resistance story from the perspective of "national unity." Two scenes give special prominence to the bicycle: In the first, our two heroes have stolen two bikes, but the conductor soon discovers he is unable to ride his. He then compels the painter to give him his bicycle, which proves to be easier to ride. The painter accepts the exchange in spite of the blatant contempt shown by the conductor. Metaphorically speaking, the humble man aims to save them both and France into the bargain.[22]

In the second scene, the viewer sees a German night patrol in a village. One of the heroes' bicycles has been abandoned in a nearby street and a wheel is still turning, a sure sign of recent use that might betray their presence. The patrol takes no notice. The sequence is a replay, with the audience smiling in approval of a familiar moment in French Resistance stories. And indeed, there were numerous such episodes featuring a bicycle in the true history of the Resistance; these might be transmitting clandestine messages or Resistance members narrowly escaping on bikes. Countless moments like these featuring bicycles were cited in authentic reports of Resistance actions (court depositions, the works of historians, war memoirs).[23] In other cases, a bicycle was necessary to go and alert people threatened with arrest. In the following example, taken from an interview with S., a Jewish family who survived the war hidden in the Massif Central describes how they were saved by the caretaker of their building. They are very conscious that this gesture entailed a perilous ride through mountainous territory, performed by a man unaccustomed to such physical effort:

"We lived in Lyons, where we had taken refuge. We left for our vacation in the Massif Central. The caretaker at our building rode his bike all the way out to tell us not to go back: they had come to arrest us."

The two situations in the film are a sort of reenactment, in a familiar and burlesque way, of such all too real moments of utter tension. The bicycle as a prop is a significant link between a tragic past and the embellished memory of reality.

Ordinary Epics

Along with the Resistance and the black market, the bicycle is also associated with another symbolic wartime event—defeat. What stands out from accounts of the defeat is the technological lopsidedness between fleeing civilians and the army of conquerors. These tales of the "*debacle*" are probably the crudest as well as the ones least prone to euphemism. The defeat of the French army suddenly drove into the streets civilians terrified by aircraft and recollections of World War I. Roads were cluttered with trucks, cars, horse-drawn carriages, and bikes. While cars soon ran out of petrol and became easy targets for machine-gun fire from the air, bicycles managed to bypass the roadblocks.[24] Those road trips involved extraordinary physical efforts, like riding hundreds of kilometers on unsuitable vehicles. Witnesses emphasized the huge technological difference between French soldiers seen escaping by bike and the German army driving cars or tanks. The opposite would become reality when the advancing American and British Armies compelled German soldiers to steal bikes for a quicker getaway; in the eyes of witnesses, this was the most obvious sign of a reversal of hierarchy between the warring powers.[25] Similar tales exist in other European countries like the Netherlands.

The fundamental reason for a reversal of memories linked with the bicycle (memories of painful events turning into good memories) seems to lie in the fact that in wartime, only the bicycle could provide a mobility that was largely uncontrolled by the state or the occupiers. It was also associated with moments of obvious solidarity within families or friends, whether or not those relationships were long-standing or recently found through unprecedented circumstances.[26] Acts of solidarity ensured the survival of the group and gave rise to feelings of love and pride. The long, arduous bicycle rides are remembered in family's tales as exceptional episodes, when husbands, wives, siblings, and lovers all became helpmates to save individuals from a tragic collective fate. At times when neither the state, nor the army, nor the institutions could assume their role anymore, friends and family stepped in.

This is how the O. family remembers one of those heroic rides: Sometime around the landing of the Allied Forces in Normandy, L. O.'s fiancé undertook a bicycle ride from Brittany to Paris, carrying the disassembled pieces of his intended's bike. He was going to collect L. O., a student in Paris, and bring her back home. Her three hundred–kilometer ride back on a badly reassembled vehicle was extremely taxing and never forgotten.[27]

The bicycle also allowed schoolchildren, temporarily exiled to safer places in the countryside, to visit their families still living at home. It helped city people to flee the bombings faster. And, repeatedly, the bicycle proved essential in the search for food that drove famished city dwellers to the countryside. Countless accounts of such ventures can also be read in memoirs published by their authors or by smaller publishers. The elements are largely the same: a bike ride by night to friendly or well-known farms; the ride back, carrying in the dark such prohibited goods as meat or butter; and the difficulty of securing unwieldy packages onto the bikes with no proper string, and of course no elastic bands.[28]

Altered Memory

Inscribing wartime bicycle episodes in the collective memory is essentially paradoxical. The vehicle may revive visual memories of grievous moments, but the reconstructed memory developed around the war as a whole tends to wrap those memories in another narrative, pointing to the bike as the one and only means of transport left. Thus the French elevated the bicycle into being one of the props that helped regain the dignity (the Resistance) or assert the companionship (through love, family, solidarity) of civilians confronted with state inadequacy or violence. For example, we see that one of the most widely read novels in the 1960s Régine Deforges's *The Blue Bicycle*, is set in wartime.[29]

Poverty, Exclusion, Marginality

The period between the late 1950s and the early 1980s marked a change in the bicycle's status in French society. The car became an ordinary consumer good: it was now the normal mode of transport for long-distance leisure trips and for commuting from home to work in provincial town and suburban and rural areas. During this period, using one's bike on daily trips meant belonging to a declining or marginal social group, estranged from the general progress and mechanization that characterized France during "the glorious thirty years."[30]

The postman in Jacques Tati's film, *Jour de fête*, typically epitomizes this situation.[31] Postmen were equipped with bicycles from the very beginning of the twentieth century, but in the early 1960s, the government wanted to give them cars. Faced with the "Americanization" of the postal service, manifested by the adoption of various unfamiliar forms of transport, Jacques Tati's postman is seen trying to rival the newcomers in speed, riding his old bike. He is a symbol of the hopeless, slightly ridiculous—though endearing—resistance of social groups left behind by social and technological change.

The figure of the countrywoman on a bike is equally symbolic. We see a character from the press illustrations of the time, akin to caricature. Mam' Goudig is a peasant woman whose name indicates she belongs to a linguistic minority.[32] When she is seen riding a bike, wearing her Breton costume and monumental headdress, her figure elicits a smile. Here the bike is linked with a character under triple domination: as a woman, a peasant, and a member of a linguistic minority. The commercial success of the character, replicated on all sorts of objects (postcards, mugs, tableware, and tourist souvenirs) reveals Mam' Goudig as a cliché belonging to "popular" culture.

Private memories also indicate that contemporaries were conscious of the social hierarchies now attached to the bicycle. Firstly, the stories about the Occupation all show that the families who owned a car wanted to get back to using it as soon as possible. The bicycle was but an unavoidable digression.[33] Besides, people who were too poor or too young to get a car in the 1960s turned to a hybrid vehicle, the motorized bicycle, or moped. These were less expensive than motorbikes and offered an alternative solution. The vélosolex or mobylettes manufactured by Peugeot (with the highest sales by far in France) were also unmistakably classified within a technical and social hierarchy.

The vélosolex—a moped, written in French with a small *v*, one of those items whose trade name has become a common term—looks decidedly like a bicycle. It has a two-stroke engine and a transmission system based on a roller bearing on the front tire to drive the wheel. In rainy weather, there is a near total lack of power, and the machine cannot go over 30 km/h, but it is simple and robust. Five million were sold in France between 1946 and 1988.[34] By the end of the 1970s, the machine became outdated and deemed obsolete ("solexes are for priests").

In the 1970s, the vélosolex was primarily used by college students, and indeed by the clergy. At least these are the two groups mentioned in middle- and upper-class people's memoirs. The vélosolex, a moped, was ridden by others in the countryside. A closer examination of private memoirs revealed that the vélosolex, considered as a kind of bicycle, could be subtly linked to subjugated positions. In a largely autobiographical story, Françoise Guérin recalls her grandfather's first purchase of a vélosolex. A retired former mason, the old man spent his entire monthly pension on a solex. The narrative subtly describes the naive pride of the old man, who can only afford a vélosolex: "The vélosolex was not a recent invention, but my grandfather's, bought second-hand, was the nicest one on earth. Slim and dignified, it had modest handlebars, a cushioned saddle and a large luggage-carrier with two side-bags fastened with the same kind of silvery buckle that my school-bag had. The name Vélo Solex, written in full, was embossed on the engine."[35]

The vélosolex, a souped-up bicycle, was also the vehicle of the less well-off, not to say poor. Both the car and the vélosolex were distinct social indicators; this much we can glean from the memoirs of Roland Bourdais. A soldier at the time, when on leave, he arrived by train at the station of the nearest provincial town at one in the morning. A young woman is there in her car, waiting for another soldier, a doctor. Roland Bourdais would never dare ask these elegant young people, his neighbors in fact, for a lift. "So, like many soldiers, I would walk to the Café de la Gare to collect my solex, and then ride along in the night on my two-wheeled vehicle all the way to Mayenne, thirty kilometers away."[36]

Using a vélosolex is also linked to gender. Young women from the country do not drive cars. Josepha Coulon assigns a large section in her autobiography to what she calls her "solex memories," because, she writes, "as I used it from 1957 to 1963, I have plenty to say about it." A young secretary in a small town in western France, she rode home every weekend to her parents' place in the countryside, quite a distance from where she worked. The solex, she says, has been a desirable object in all classes of society: "that black bicycle running on only one liter of solexine for a hundred kilometers required a skillful start, as you had to manage the 'solex handles.' It greatly helped by saving your breath riding up slopes, even though you had to pedal a bit." If it happened to break down, you had to go into town, leaving before daybreak on Monday: "then, pedaling in the dark on my bike with no light, I did not feel too safe riding through the Soeuvres woods."

Another group felt left out because its members were refused access to the car: teenagers who had to be eighteen to sit for their driving test and obtain a driver's license. The mobylette became a kind of initiation rite in French society. Being relegated to the status of cyclist (for urban rides) was in some way being denied the full status of a young adult. The only acceptable use of a bicycle for a teenager was racing. For everyday use, the possession of a mobylette, which one was allowed to drive from the age of sixteen, gave young people the opportunity to mimic owning a car: modifying the engine, for instance, or customizing the body. Being offered a moped as a present from the family, often related to some school success, had a distinct feeling of initiation, while clearly reinforcing the receiver's teenage status.[37]

Conclusion

What conclusions can be drawn from this brief survey, spanning several generations, of the way memories were created that were somehow linked with the bicycle?

I have highlighted the fact that the usual sources are somewhat biased. Archives and libraries have primarily collected magazines, posters, and documents with the intention of promoting the use of the bicycle. Such sources naturally offer predominantly positive images and stories of bikes. Nevertheless, an analysis of memories rooted in the war and occupation indicates that the effect of memory also introduces filters that are apt to transform negative circumstances into positive recollections.

My analysis of authentic life stories, which I highlight in the latter part of this essay, points us in two directions. First, in autobiographical reports we can speak about circumstances of poverty, exclusion, and social domination, and through these circumstances affecting ordinary lives, relate to issues linked to the social history of technology. Second, such narratives reveal the roots of significant attitudes toward the bicycle.

Immediately after the war, all the former owners of cars returned to this form of transport as soon as possible. From the 1960s onward, motorized bicycles were the everyday vehicles of choice for all those who could afford them. This choice seems to stress the enduring difficulty of dissociating the use of the bike (for getting to work) from its early context, both rural and lower class. The reintroduction (or reimposition) of the urban use of the bicycle in France around the late 1990s, therefore, necessitated a serious media campaign in order to imbue the project with new representations, like those of young urbanites belonging to the new intellectual or liberal professions.

Seen in a historical perspective, the representations linked to the use of the bicycle for daily transport are diverse as they refer to specific periods in time. These representations are part of a collective memory, both through family histories and the national collective story. Products of popular culture—songs, plays, or films—keep alive and reinforce those representations, as they deliberately make use of stereotypes and references commonly shared by French audiences.

On the whole, in early twenty-first-century France, these representations were perceived as ambiguous. Because "the bike" brought to mind the warm sense of fellowship and solidarity prevailing within families and circles of friends developed in dangerous war times, it assumed a positive aura. On the other hand, once replaced in its unglamorous use for daily transport, especially to get to some equally unremarkable workplaces, as was the case around the 1960s, the bicycle has come to be associated with social groups left behind by the ongoing socioeconomic growth.

Reintroducing the bicycle as an exalted form of urban transport requires a serious effort from authorities committed to the task. Aiming at new consumers, this modern communication would require the transformation of the vehicle's old image. We would have to invest in a lingering sympathy for the object, while toning down its negative associations. This is precisely the method used in 2000 by the designers of the posters created for the launch of the Vélib, a new self-service bike-rental system in Paris. What catches your eye in the Vélib poster is a young upper-middle-class woman on a bicycle; she is neither working class nor from the countryside. The Paris authorities then thought up a number of accessories, such as a front wire basket to hold a purse or handbag. Obviously the rider cannot be burdened with heavy packages, cumbersome shopping bags, or children, but smaller items are easy to carry.

In today's urban environment, where users can choose between various forms of transport, a successful reimaging that calls to mind gratifying representations of the bicycle could impact the prospective user or buyer's choice. Where mobility is concerned, the choices may be limited by purely rational criteria. But one must not neglect the insidious force of the user's self-image.

Catherine Bertho Lavenir, professor of contemporary history at the Sorbonne (Université Sorbonne Nouvelle), Department of Arts and Medias, is today director of the education services in the French Region of La Martinique. The author of *La Roue et le stylo comment nous sommes devenus touristes* (1999), a book on automobile mobility and the building of a travel culture, she specialized in history of mobility and interested herself in the social and cultural history of bicycle. She published *Voyages à vélo* (2011), a cultural history of French bicycle. Her last publication is "Grand tourisme automobile et identité européenne 1915–1970," in Marc Gigase, Cédric Humair, and Laurent Tissot's *Le tourisme comme facteur de transformations économiques, techniques et sociales (XIXe–XXe siècles)*.

Notes

1. *La demande de trafic routier. Relever le Défi. Rapport de l'OCDE* (OCDE Publishing, 2002); Timothy Beatley, *Green Urbanism: Learning from European Cities* (Washington, DC, 2000).
2. Mathieu Flonneau, *Parcourir et gérer la rue à l'époque contemporaine, Pouvoirs, pratiques, représentations* (Paris, 2008); Mathieu Flonneau and Antoine Prost, *Paris et l'automobile un siècle de passions* (Paris, 2005).
3. For example, "*La bicyclette*," by Yves Montand in 1968. The author of the lyrics, Pierre Barouh, writes: "'*La bicyclette*' comes directly from my childhood in Vendée.... These songs are not only popular hits; they also are part of the popular memory. Few people know that I wrote [this song]"; www.jechantemagazine.com.
4. For a comparison with Finland, see Tiina Männistö-Funk, "The Prime, Decline and Recalling of Rural Cycling: Finnish Cycling in the 1920s and 1930s Remembered in 1971–1972," *Transfers* 2, no 2 (2012): 49–69.
5. Maurice Halbwachs, *Les cadres sociaux de la mémoire* (Paris, 1925), 211; Nicolas Roussiau and Christine Bonardi, *Les représentations sociales, Etat des lieux et perspectives* (Brussels, 2001).

6. Henri Rousso, *Les Années noires – Vivre sous l'Occupation* (Paris, 2006). More than 7 million licences were sold for bicycles in 1940 and almost 11 million in 1942.
7. Anne-Françoise Garçon, ed., *L'Automobile, son monde et ses réseaux* (Rennes, 1998), 162.
8. Interview A. L. (72 in 2011). A press photo from Roger Viollet's collection illustrates such inventive contraptions. It also bears witness to this kind of downslide in social status.
9. Interview A. L. (72 in 2011): "My father worked in an office. We lived in an eastern suburb of Paris. He had to buy a bicycle to get to work in the west of Paris"; J.-C. La. (78 in 2011): "My father was the director of a factory near Saint-Ouen. We lived in Versailles. In 1944, when the bombings intensified, he took to cycling to his office, leaving very early in the morning. It was about 25 km away."
10. Similarly pernicious images were produced during the German occupation of the Netherlands. The Nazi propaganda film *Warsaw Ghetto,* shot two months before the mass exterminations, staged scenes of Polish Jews being chauffeured in taxi-bikes by non-Jews. See documentary by Yael Heronski, *A Film Unfinished* (2010).
11. Cordula de Marx, *Die französische Wochenzeitschrift 'L'Illustration' während der Zeit der deutschen Besatzung 1940–1944* (dissertation, University of Würzburg, 1993).
12. René Baschet, "Ré-inventer," *L'Illustration, no. spécial l'Automobile* (1941): 5–6.
13. Pascal Ory, *La France allemande, 1933–1945* (Paris, 1977).
14. "Generally bike-trailers made of board, planks, and tin, which serve as goods-carriers during the week and pick up passengers on a Sunday ... they are exactly the same trailers that could be hitched to the handy, living steed of steel, which used to be a bicycle." Baschet, "Re-inventer."
15. Ibid., 6.
16. François Toché, "Cyclotourism 1941," *L'Illustration, n° spécial L'Automobile* (1941).
17. "When petrol sources suddenly dried up last year, we went back to the bicycle as fast as we could. Shops were instantly relieved of their stocks, factories were overwhelmed by orders. We started searching in barns, lofts, second-hand and antique shops. Today an immense army of cyclists sets off every morning. Offices and factories have had to enlarge parking areas for their employees' bikes. Every residential block now has its own cycle shed in the stairwell. Bikes reach all the way to the top and out to the back whenever a tenant entertains friends after dinner." Toché, "Cyclotourism."
18. Ibid. "The car has disappeared from our lives for the time being, requisitioned, sold, abandoned along the evacuation roads or yet again sent into limbo: a dead garage. Let's be honest. Did we really know how to benefit from such a wonderful means of escape? Had we not too often become obsessed by performance, near robots on the road?"
19. Ibid. "We marvel at the discovery of the riches and beauties of that magnificent park, the land of France."
20. Jean Baronnet, *Les Parisiens sous l'occupation. Les photographies en couleur d'André Zucca* (Paris, 2008), 176.
21. Larry Portis, "L'état dans la tête et les pieds dans le plat." Hiérarchie et autorité dans les films de Louis de Funès, *L'Homme et la société* 4, no. 154 (2004): 31–50.
22. Stanislas: Vous allez garder le mien qui ne marche pas et moi, je vais prendre le vôtre. Allez hop! ("You take mine which doesn't work and I'll take yours. Off we go!")
    Augustin: Mais dites donc, ça fait deux fois que vous me faites ça. ("But that's twice you have done this to me.")

Stanislas: Oui. ("Yep.")

Augustin: Vous m'avez déjà pris mes chaussures, maintenant mon vélo. ("You've already taken my shoes, now my bike.")

Stanislas: Eh bien alors c'est normal, non? ("Well now, there's nothing wrong with that, is there?")

La Grande Vadrouille, Georges et André Tabet, 1966, Bourvil et Louis de Funès.

23. Site "Pour que la France vive.Histoires-libres.com.... Témoignage de Mr Jean Letouzey (né en 1921)": "I decided to return to Tonchebray. My sister Suzanne, a little bit younger, met me in Ceaucé. As I had to bring my weapons, I hid them in the bags attached to my sister's bicycle, buried under clothes. She wasn't aware of this. We were on the road. Everything was fine, then suddenly I had the worst scare in my life, especially fearing for my sister."
24. Marc Bloch, *L'Étrange Défaite* (Paris, 1990).
25. http://jacquotboileaualain.over-blog.com/article-la-liberation-de-champagney-1-97101919.html.
26. Remarkably in the Netherlands, the Germans confiscated bicycles on a grand scale, yet there we find a similar celebration of the bicycle as a means of freedom/mobility to search for food during times of scarcity. Personal communication with Ruth Oldenziel, August 2012.
27. Interview D. O. (82 in 2011).
28. Interview F. Le. (59 in 2011, daughter). See also Paul Vannier, *Un si bel été. Memoires de la drôle de guerre* (Paris, 2007), xiii.
29. Regine Desforges, *La bicyclette bleue* (Paris, 1981).
30. Jean Fourastié, *Les Trente Glorieuses ou la révolution invisible de 1946 à 1975* (Paris, 1979).
31. Jacques Tati, *Jour de fête,* 1947.
32. Jean-Paul David created this character of a Breton peasant woman ten years ago. Already present in the 1980s, the stereotype generated images thirty years later.
33. Daniel Coulaud, *Ville, automobile et modes de vie* (Paris, 2010), 97.
34. Patrice Huerre, Martine Pagan-Reymond, and Jean-Michel Reymond, *L'adolescence n'existe pas. Une histoire de la jeunesse* (PUF, 2003 [1990]), 287.
35. Françoise Guerin. Autobiographie-Blogspot. https://books.google.fr/books?id=tP7RdzLWQF0C&pg=PA86&dq=inauthor:%22Fran%C3%A7oise+Gu%C3%A9rin%22+velosolex&hl=fr&sa=X&ved=0CBwQ6AEwAGoVChMI7q67tsfkxwIVBX0aCh2G9ASI#v=onepage&q=inauthor%3A%22Fran%C3%A7oise%20Gu%C3%A9rin%22%20velosolex&f=false
36. Josepha Coulon, *Mémoire ma complice* (Paris, 2002), 77.
37. Gilbert, Le Bleu Marine nationale, autobiographie, http://lebleumarinenationaleautobiographie.blogspot.com, 17 August 2011.

 CHAPTER 4

Who Pays, Who Benefits?
Bicycle Taxes as Policy Tool, 1890–2012

Adri Albert de la Bruhèze and Ruth Oldenziel

Vittorio De Sica's film *The Bicycle Thief* (*Ladri di Bicicletta*, 1948) shows the iconic place of the bicycle in postwar Italy by narrating the wrenching story of a poor worker who loses his bicycle—and thus his ability to land a job.[1] The movie brilliantly captures the class politics surrounding cycling, an issue that continues to affect our current sustainable mobility debates. This chapter will focus on one aspect of this issue, namely bicycle taxes. Over a century ago, the debate over bicycle taxes was connected to creating infrastructures for the new mobility of bicycles and automobiles: who would benefit, use, and pay for new road systems? The answers to these questions hardened into blueprints of class-based road building that are still with us today. The bicycle tax debate has reemerged in full force again in a seemingly radically different cultural frame of reference. No longer seen as a working-class vehicle, the bicycle has been reframed within the overall narrative of solving traffic congestion and reducing CO_2 emissions. There is a general consensus that cycling is a healthy, fast, clean, and sustainable mode of urban transportation. What has changed dramatically since De Sica's movie is the recent reframing of the bicycle as a vehicle of choice for morally superior green citizens in the Western world. Today's discussions focus on whether bicycle taxes should be levied to cover bicycle-related costs: building bicycle lanes, maintaining infrastructure, combating theft, creating bicycle-related employment for the long-term unemployed, and enforcing traffic laws.[2] Urban politicians have discovered that sponsoring bicycle-share programs is a good political decision.

The debates about reintroducing the bicycle tax are about more than just raising money, however. The proposals often reopen fundamental questions about who has the right to the roads that date back to class-based debates in the 1920s and 1930s, if not earlier. Apart from anecdotal accounts, there is little historical scholarship on the topic.[3] Despite the radical shift in cultural framing, the current bicycle tax debate—routinely couched in neutral terms of road safety, urban planning, and traffic engineering—shows striking similarities to the debates of a century ago. In questioning who the "legitimate" road users are and who should benefit from taxes, the current bicycle tax issue also reveals new challenges: what is the cost of the transition into sustainable cities, and how much are citizens and political leaders willing to discuss these changes? Arguments of the past thus continue to shape the current debates. In this chapter, we argue that even when cultural frames of reference radically change—from De Sica's proletarian protagonist to the moral protagonist of the green movement—old class-based biases persist and negatively impact sustainability policies.

Public Good and Bicycle Taxation, 1890–1945

Taxation is never neutral. It is a deeply politically charged policy that expresses ideas about the common good as well as power relations.[4] Any kind of tax is guided by two economic principles: who can pay and who stands to benefit. The ability-to-pay principle mandates that wealthy citizens and profit-making businesses carry the largest tax burden simply because they can—and should—contribute to the public good. The benefit principle considers whether taxpayers benefit, either individually or as a special interest group, from specific government services and should therefore pay for them.[5]

The history of bicycle taxes can be divided into three phases. Each highlights what was considered the public good; each also shows the contests over who should pay for it. During the first phase (1880s to 1920s), bicycles were luxury items that could be afforded only by the well-to-do. Local and national governments taxed bicycles in order to raise general revenues. In this liberal laissez-faire period, the nation-state sponsored few roads directly. Filling the gap, highly organized, mostly well-to-do cyclists privately financed the building or improvement of separate roads through fundraising, donations, and club dues. During this period, they also sought to redirect and earmark bicycle taxes for networks of dedicated bicycle lanes separate from the main arteries because cyclists needed smoother road surfaces. To gain the support of rural groups, anti–bicycle tax activists began to favor investing in good roads for all instead of in separate cycling infrastructures.

In the second phase, 1920–1940, the price of bicycles dropped dramatically, bringing these modern vehicles within reach of many more citizens. This was also the moment when the nation-state began to create national infrastructures. In this period, cyclists, mostly from working-class backgrounds, found themselves forced to pay taxes without benefitting from the modern road infrastructures and without enough political resources to resist the taxes. As we will argue, working-class cyclists were effectively excluded from the new definition of the public good that ignored their individual mobility needs as a matter of principle.

In the 1970s, after a four-decade hiatus, bicycle taxes reappeared on the agenda, but in a radically changed discursive context. In this third period, bicycle taxes steered clear of class controversies, were embedded in sustainability visions, and were seen as a potential tool to gain political leverage for encouraging cycling and financing bicycle infrastructures. The debates, however, echo earlier class-based contests that restricted and marginalized cyclists.

Debating Bicycle Infrastructure as a Public Good, 1890–1920

From the 1880s to the 1890s, state authorities levied luxury taxes on bicycles simply because the well-to-do could afford to bear the financial burden in countries from France, Italy, Belgium, and the Netherlands to Spain, Portugal, Switzerland, and Austria.[6] In this first phase, the bourgeoisie, using their bicycles to tour the countryside and urban parks, paid bicycle taxes, registration fees, riding permits, and other costs on top of the purchasing price. These costs for users effectively prevented workers from buying the novel consumer goods that symbolized true modernity and personal freedom. In 1893, France and Belgium were the first countries to introduce a national luxury tax on bicycles when the number of prosperous cyclists peaked. Italy soon followed in 1897.[7] At the same time, middle-class cyclists organized in clubs and used membership fees and club donations to fund the improvement of roads that were not yet adapted to the needs of such a modern vehicle. Technically, bicycles had already fully developed as an innovation, but the bottleneck to their full potential was the many unpaved and bumpy roads—and national states and provinces were not in the business of improving them.[8]

Cyclists sought to sponsor better roads for their new mobility mode through private funds, by redirecting bicycle tax revenues, by seeking access to public funds, or a combination of such strategies. As James Longhurst has documented for the United States, Charles T. Raymond, an avid American cyclist and successful businessman, was a driving force in the efforts to fund dedicated bicycle infrastructures. In some US cities (Minneapolis, St. Paul, Rochester, and Portland) and states (New York, Maryland, Ohio, Florida, and Minnesota), the US sidepath movement succeeded in getting laws to tax bicycle owners for this dedicated purpose.[9] It was an international phenomenon. In Lübeck, Germany, cyclists funded the urban lanes through their bicycle luxury taxes between 1900 and 1919.[10] In Magdeburg, the influential municipal engineer and cyclist Carl Henneking was able to get 400 kilometers of dedicated bicycle lanes built in and around his city by 1929. They were financed through a combination of private club donations and local public funds.[11] In the Netherlands, supported by the national bicycle organization ANWB, local bicycle chapters financed and built rural bicycle lanes for touristic purposes.[12]

When proposing bicycle taxes, authorities tried to convince cyclists that they stood to benefit, promising that the revenue would be used to fund separate bicycle lanes. In the 1890s, authorities employed such "benefit" arguments during the tax debates in Austria, the Netherlands, Great Britain, France, and Belgium.[13] In practice, the benefits were often sacrificed for more politically expedient issues. In Belgium, for example, the first bicycle taxation of 1892 and the first national traffic law of 1899 were initiated by the *Ligue Nationale pour l'amelioration des Routes*, in which cyclists and motorists collaborated. The Ligue succeeded in lobbying for taxes for more accessible ("good") roads (*Maak goede wegen*!) and bicycle lanes in five provinces (Liège, Antwerp, Hainaut, East-Flanders, and West-Flanders). These lanes were never built, however: the revenues were diverted to the Poor Fund instead.[14]

Increasingly, cycling advocates opposed bicycle taxes altogether on the grounds that bicycles had become so integrated into everyday life that they should be considered extensions of the body and therefore be tax exempt. They questioned the idea that bicycles were luxury items.[15] There was some basis for their challenge. All over the industrializing world, mass production had led to a steep decline in prices, bringing bicycles within reach of the lower middle classes in less than a decade.[16] When the city council of Graz, the heart of Austria's bicycle industry, proposed bicycle taxes in 1894 (and again in 1898) to fund an expensive theater that mostly benefited the local elite, a broad-based grassroots movement challenged the assumption that only the rich would be affected by the bicycle tax. In their view, the tax was socially unacceptable because government employees, like the military, police, and post office workers, as well as small entrepreneurs, farmers, and most workers, used bicycles daily for their work.[17] The movement also challenged the bicycle tax because it singled out cyclists, while pedestrians and carriages were not charged for using roads. Moreover, separate bicycle lanes were more to protect careless jaywalkers than to benefit cyclists, they argued.[18] The bicycle activists from Graz supported their arguments with international evidence. After conducting a survey among four hundred sister bicycle clubs in cities throughout the Austrian-Hungarian Empire, Bosnia-Herzegovina, the German Empire, Belgium, France, Italy, the Netherlands, and Switzerland, the Austrian cycling activists concluded that the local authorities' push to tax bicycles was outdated. The bicycle had become a necessity for many, cycling a general practice, and its taxation a social injustice.[19] The Austrian cycling coalition succeeded in defeating the measure.

Cycling advocates thus began to question luxury taxes as an easy way for state authorities to fill their coffers. While resisting the taxes, cycling activists also tried to redirect the bicycle revenues to infrastructures dedicated to cyclists, such as their better-roads initiatives. Many other urban-based club leaders concluded that it made more sense to invest in better general roads that would benefit all than in separate bicycle infrastructures, and they made common cause with rural interests that had resisted cyclists' lobbying efforts at first. The coalition led to many national "good roads" associations that sought access to public funds.[20]

Indeed, in the process of resisting the luxury tax and in redirecting revenues for their cause, cycling activists were redefining the terms of public good for the modern mobility of cycling: roads were to be a service accessible to all whether one paid taxes or not. How those roads were to be designed and who would be a legitimate road user became a hotly contested issue in the years to follow.

Roads as a Class-Based Public Good, 1920–1940

On the grounds that bicycles were no longer a luxury, bicycle taxes were phased out in many nations. Nevertheless, bicycle tax proposals were soon on the political agenda again after the First World War. This second phase (1920–1940) of the bicycle tax debate was based on the opposite rationale: Governments saw new opportunities to tax bicycles as an easy revenue-raising measure precisely because the mass-produced bicycles were no longer a luxury and so many citizens—first small entrepreneurs and civil servants, then (male) workers—used the bicycle to commute. The political environment of the bicycle taxes, however, had changed radically. The tax debates coincided with contests over what constituted public good at a time when state authorities were fashioning new public roles in shaping infrastructures, as James Longhurst argued so well in his case study of the United States.[21] In the industrializing world, states became more directly involved in large infrastructures like mobility networks, electrical grids, and food systems because the scope of those infrastructures posed challenges that local communities could no longer handle, coordinate, or fund.[22] States also needed funds for such large projects. The cycling working class offered an attractive fundraising opportunity.

When bicycle taxes were proposed again, they became mired in class politics in a decade of war and revolution. Labor parties entered the political arena in the many national and local elections after 1919 amid the upheaval of the First World War. While the antitax argument—bicycles were no longer a luxury, but a basic need—was advanced once again, the argument fell on deaf political ears. From the United States to Britain, the Netherlands, France, and Austria, the bicycle tax debate moved to state and national parliaments, where socialists sought to represent the millions of working-class cyclists and resisted the reintroduction of bicycle taxes. In many instances, the representatives of labor were often outright defeated when the taxes came up for a vote. Significantly, the bourgeois support was missing. The middle classes, who had been so successful in their advocacy earlier when bicycles were still high status, now moved their allegiance to the new high-status vehicle of the day: the automobile. In the second period, automobile advocates, local authorities, and traffic engineering experts successfully contested working-class individual mobility needs. They framed bicycles as a utilitarian, slow, old-fashioned, dangerous form of transport.[23] The debate became a struggle over the meaning of and access to roads as a public good.

Well-to-do car advocates, such as automobile clubs, and road builders succeeded in arguing that this new mode of transportation would inevitably prevail, and that dedicated road infrastructural support for it would be in the nation's best interest. This was remarkable: in the 1930s, by far the highest proportion of traffic was made up of pedestrians and cyclists—not motorists.

Figure 4.1. Bicycles' Share of the Total Number of Car, Public Transport, Bicycle, and Moped Trips in Nine European Cities, 1920–1995.

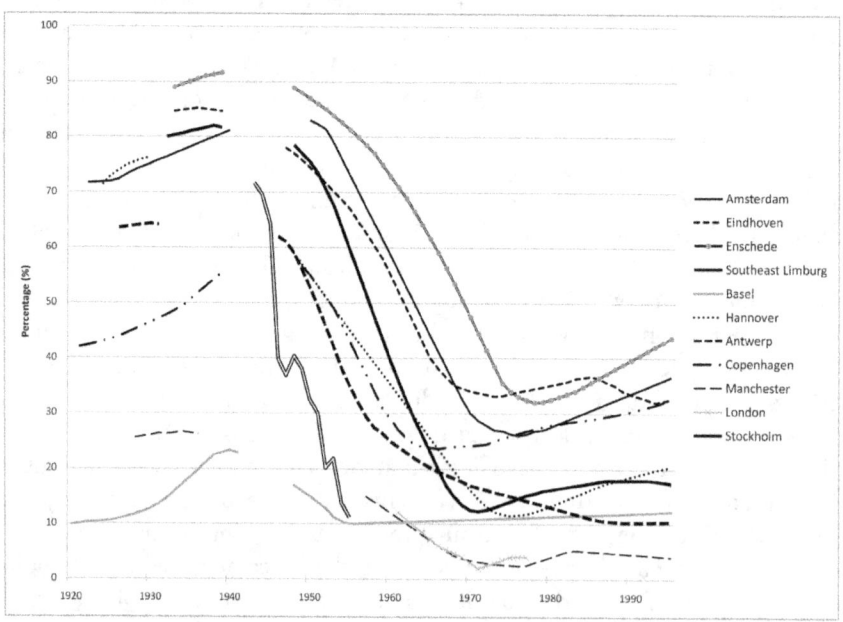

From the First World War until the early 1960s, the bicycle was a dominant form of transportation. This chart compares bicycle use to cars, buses, streetcars, and moped in ten urban areas in six countries, showing cyclists dominated the streets.

Source: Oldenziel and Albert de la Bruhèze, *Contested Spaces*, 33. The graph is an extended version of Albert de la Bruhèze and Veraart, *Fietsverkeer*, 14. The trend lines are based on modal splits and traffic-count data; where no sound assessment could be made, the trend line has been omitted.

And the number of cyclists was booming. Yet, automobiles and large interstate highways became state priorities, rather than cycling infrastructures. Politically—and literally, as shown by mounting accident rates—upper-class motorists and working-class cyclists were on a collision course.[24]

When it came to building new road infrastructures that were in the public interest and the question of who would foot the bill, experts and policy makers sought to diffuse the growing class-based tensions over what constituted the public good. Engineers were crucial in helping to articulate the national importance of car-governed infrastructures as the way of the future. They ignored urban working-class mobility needs from the start. Together with upper-class motorists and politicians, state engineers helped remove the decision-making process over taxes and their allocations from the political arena to the drafting departments of state agencies by advancing technical arguments.[25] In the policy discussions on whether roads should remain a space where all traffic shared the road or a monofunctional space privileging motorized transit, the latter view quickly came to dominate engineering circles. The vision privileged and reframed cars as the premier public good. Between 1910 and 1926, internationally networked engineers and policy makers argued that the best way to make space for fast traffic was to exclude pedestrians, cyclists, and other nonmotorized traffic—now framed as slow and old fashioned.[26] For this argument to stick, they also had to show that the relatively few automobiles on the road—on average no more than 4 percent of the total traffic in these decades—were the vehicles of the future and that the numerous cyclists were not only increasingly old-fashioned, but also slow, irresponsible, and dangerous road users.

In the years following the 1926 international Road Congress in Milan, when a standard for traffic counting was established, traffic counters concluded that chaos reigned in the streets.[27] To solve the problem, traffic engineers with both local and international experience articulated that bicycle lanes should not be built for the comfort of cyclists, but only to ensure motorists would no longer be hindered by slow traffic. Italo Bonardi, MP and member of the Italian Touring Club, summarized the international trend. He admitted in 1938 that cycling lanes were to be built to "liberate motorists of cyclists and cyclists from motorists." He also reminded his fellow politicians to stop complaining about cyclists, because they—over 4 million of them—filled the government's coffers with their taxes.[28] Notwithstanding Bonardi's reminder, the vision of segregation was further hardened through restrictive legal measures against cyclists and pedestrians that gave motorists the freedom to move ahead.[29] To further the cause, the United States, the Netherlands, Italy, and Germany carried out the first experiments with national highways (1916), *rijkswegen* (1927), *autostradas* (1929), and *Autobahnen* (1933). These roads served urban elites, agricultural lobbies, and military interests.[30] And as a matter of principle, the national road-building plans of the 1930s excluded cyclists, pedestrians, horses, and carts by law and engineering design. This was also de facto the case for the older roads that were upgraded for the new mobility: motorized traffic became the standard.[31] Traffic engineers, urban planners, and economic statisticians played a crucial role in devising how to exclude traffic participants, now designated as slow and dangerous.[32]

The case of the 1924 Dutch bicycle demonstrates how a class-based definition of the public good was reframed and executed through tax policy, and the far-reaching consequences this had. Originally, the bicycle tax reintroduced in 1924 was meant to be a temporary measure to raise revenues. It generated a storm of protests from working-class parties, civil servants, and the tourist organization (ANWB)—to no avail. In the first year the measure went into effect, the Ministry of Finance was pleased to discover that 1.7 million citizens owned bicycles instead of the projected 1 million.[33] The state could hardly resist making the temporary revenue-raising measure permanent. Under political pressure, the bicycle tax revenues were redirected two years later; now they flowed—together with car-taxation revenues—into the Road Fund.

The shift from a general tax (benefitting all) to a dedicated one (benefitting those who paid it) was neither politically neutral nor insignificant. An instrumental role was played by the automotive lobby, which wanted to fund an exceedingly ambitious road construction and retrofitting plan. The lobby included civil society groups (the Dutch tourist organization ANWB and the Royal Dutch Automobile Club KNAC) and commercial interests (Union for Business Car Owners BBN, Royal Dutch Road Transport Association KNWV, and the Association of Chambers of Commerce), and three major construction companies (Dutch Concrete Society, Dutch Basalt Company, and NV Bitumenweg). They lobbied—and got—a road fund to build car-based roads and the car-only highways that would be connected to a national network to serve long-distance motorized traffic in the distant future. It was a massive national undertaking in times of economic crisis and impending war. The costs of improving or replacing existing national and provincial highways as well as constructing new roads and bridges over the country's wide rivers were projected at €2.3 billion, at today's rate (300 million guilders) over thirty years. By 1940, €3 billion (380 million guilders) had been spent.[34] In the 1920s, the emerging national coalition's lobbying efforts—like those of their international counterparts—was not a response to pressing needs, but were geared toward a projected future.

Indeed, automobility was still a vision of the future rather a reality. By contrast, cyclists, most of them workers, were overwhelming the streets in ever increasing numbers in the decades between the 1920s and 1950s. Indeed, traffic statisticians initially excluded cyclists from their traffic-count sheets, but they discovered that cyclists dominated European streets. Urban cyclists often had a modal share of over 70 percent of the total number of trips taken by automobiles, public transport, and mopeds.[35] These numbers had a profound impact on the politics of the bicycle and car tax revenues.

The bicycle taxation revenues flowing into the 1926 Dutch National Road Fund benefited an exceedingly small number of motorists (less than 3 percent) and disadvantaged cyclists in several ways. Cyclists, whose numbers were ex-

Figure 4.2. Dutch Cyclists' Contribution to Interwar Road Building

Year	Number Cyclists (%)	Number* Motorists (%)	Tax** Contribution Cyclists (%)	Tax** Contribution Motorists (%)	Funds*** Allocated to Bicycles (%)	Funds*** Allocated to Cars (%) ***
General Revenues 1924–1926						
1924	95	1.7				
1925						
1926			28	72		
Road Fund 1927–1934						
1927			42	58		
1928	96	1.8	47	53		
1929	95	2.2	41	59		
1930	95	2.4	39	61		
1931	95	2.6	36	64		
1932	94	2.7	35	65	5	95
1933	94	2.8	35	65	5	95
1934	94	2.8	34	66	5	95
Traffic Fund 1935–1941						
1935	94	2.7	25	75	5	95
1936	95	2.5	29	51	5	95
1937	95	2.5	33	67	5	95
1938					5	95
1939	94	2.6			5	95
1940					5	95

Between 1924 and 1941, cyclists contributed disproportionally first to the treasury and, after 1927, to the upgrading and building of motorways.

* Excluding trucks and buses. The percentage of busses and trucks increased from 1.1% in 1927 to 1.4% in 1937, with peaks in 1932 (1.8%), 1933 (1.7%), and 1934 (1.7%).

** Bicycle and car tax contributions to the road traffic funds financed the 1927 and 1932 national road plans to build and improve car roads. Thirty percent of the bicycle tax revenues flowed into the road funds in 1927, 60% in 1927, and 90% in 1928. Since 1929, all bicycle tax revenues went into the Road Fund; since 1935 into the Traffic Fund. In 1934, the Road Fund also received 70,000 guilders from the Employment Fund, created to fight unemployment. In 1935, the traffic fund received 2.68 million guilders from the Employment Fund.

*** Percentage of (road and traffic) funds spent on constructing (national and provincial) car roads and separate bicycle paths alongside them. The government budgeted 35.5 million guilders to build cycle paths in the period 1932–1957. This 1.4 million annual budget is 5% of the total annual amount spent on roads. Unclear is how many paths were actually built.

plosively growing, carried a disproportionate financial burden for upgrading and building Dutch motorways. While the government contributed only 15 percent (59 million guilders, or roughly €456 million in today's Euros) between 1927 and 1940, some 85 percent (320 million guilders, or roughly €2.5 billion), came from automobile and bicycle taxes.[36] But automobiles made up only a small percent of the taxes; most of the funds were raised by cyclists: there were 1.7 million cyclists in 1927, increasing to a whopping 4 million in 1940.[37] What Dutch cyclists—mostly urban—contributed to car roads built outside cities and what they received in return, in terms of bicycle paths, is summarized in figure 4.2.[38]

Thus, for every tax guilder paid by cyclists, only 14 cents were used for bicycle lane construction. By contrast, for every tax guilder motorists paid, almost two were used to fund dedicated car roads. In other words, cyclists actually paid seven times more than they received in bicycle lane construction, whereas motorists paid only half of the money invested in car-governed roads. But there was more to the inequity than expressed in these figures.

Traffic experts and policy makers rallied several arguments to explain why funds should go to constructing dedicated car roads instead of structures for urban cyclists and pedestrians. In his traffic segregation report to the 1938 world road congress, R. Tijken, secretary general of the ANWB's Road Committee, distinguished four traffic zones radiating from city centers to the periphery: town center, inner periphery, outer periphery, and nonresidential area (see figure 4.3). Tijken observed that the 1937 traffic counts for the city of The Hague "shows that cyclist traffic is of a preponderating local nature," and he extended the conclusion to all Dutch cities.[39] He articulated an emerging international consensus among road advocates and state engineers that their task was to cater to nonurban areas where motor vehicles dominated. The net result of the engineers' narrow focus on the so-called nonresidential area was the exclusion of urban cyclists' and pedestrians' needs from engineering

Figure 4.3. Traffic Count by Zones (The Hague, 1937)

Zone	Cycles (%)	Motor Vehicles (%)
Town Center, 0.0 (Spui)	75	25
Inner Periphery, 2.5 km (Rijswijksscheweg)	71	29
Outer Periphery, 1.5 km (Wenkebachstraat)	66	34
Nonresidential Area, 2 km (Ypenburg)	37	63

Table distinguishing the four different zones radiating from the city of The Hague and the proportion of cyclists as rationale for excluding counting cyclists outside residential areas.

Source: Based on R. Tijken, "The Segregation of the Various Classes of Traffic on the Highway," in Second Section: User, Regulation and Administration: 4th Question, 9. Permanent International Association of Road Congresses (PIARC), VIIth Congress (The Hague, 1938).

design and allocation decisions. The rationale was also behind the 1927 and 1932 Dutch Road Plans. These plans closely resembled their international counterparts.[40]

Urban cyclists not only disproportionally contributed to the national and provincial road-building funds, but also got bicycle lanes they did not need. Riding on the wave of the controversial bicycle tax, the tourist organization (ANWB) successfully lobbied the Ministry of Transport and Water for separate tourist-oriented bike paths along the newly constructed roads in 1929.[41] While impressive at first sight, the political victory hides the class dimensions. The amendment of the first Road Plan was a political compromise. The middle-class tourist organization succeeded in creating bike paths for tourists but ignored the urban working-class cyclists (see figure 4.4).

As urbanites, working-class cyclists paid a high price in other ways as well. The first national road plan not only connected, but also crossed towns and cities, forcing streets to be widened and turning them into highways. As a consequence, cities became crowded traffic hubs, where cyclists' freedom and room to maneuver was routinely and severely curtailed to make way for motorists. This would become the focal point for urban protests in the 1960s.[42]

Figure 4.4. Cyclists Sidelined

Cyclists, whose bicycle tax paid disproportionately for the newly built highway system drafted in 1926, seen in this photograph taken at the newly built highway (Rijksweg) outside the city The Hague circa 1938 with separate bicycle path. Courtesy of the Foundation History of Technology, Eindhoven.

Workers felt the disproportionate financial burden of the tax most directly. The bicycle tax of 3 guilders, reduced to 2.50 in 1928, was still a huge sum for any working-class family.[43] Decades earlier, the French anti–bicycle tax movement had similarly argued how bicycle taxes disadvantaged the working class: in terms of the purchasing price, ordinary cyclists were disproportionally taxed compared to bourgeois motorists—4 percent against 0.95 percent. Cyclists who bought second-hand bikes paid even more (25 percent).[44] Socialists and communists led the fierce opposition. To appease the explosive issue, which only intensified during the 1930s economic depression, the Dutch government sought to alleviate the burden by exempting the unemployed: half a million jobless workers applied annually in the period 1936–1940. That exemption came at a high social price, however. The unemployed were stigmatized for all to see because the exemption forced them to carry a bicycle tax tag with a hole. And they resented the stigma, as popular songs and poems testify.[45]

Indeed, the bicycle tax became the symbol of working-class resistance to government measures during the economic crisis in both overt and covert ways. To enforce the law, the police installed checkpoints at bridges and ferries where cyclists could not avoid them. In 1933 alone, Amsterdam police fined 11,809 cyclists for failing to pay the tax, and in Rotterdam, 13,207. After 1936, the class tensions over the bicycle were less public because the courts rather than the police enforced the law. Working-class cyclists continued to suffer immensely. Those caught riding without tags were punished with a high fine of 5 guilders; those with illegal exemption tags available to the military, police officers, postal employees, and foreign diplomats were fined twice that amount. Not just the unemployed suffered; even the working poor struggled. Photographs of weekly public auctions selling confiscated bicycles in front of the Amsterdam tax office suggest that many could not afford the taxes.[46] As a result of these pressures, a black market for bicycle tags thrived. Forgeries and theft created their own niches, like the Amsterdam second-hand market Het Waterlooplein—a working-class world De Sica's drama captured so brilliantly in *The Bicycle Thief*. Not only in the Netherlands and Italy, but also in Austria the fight was bitter—and the working class lost every time.

In 1935, a long drawn-out and heated debate in the state parliament of Styria, Austria, over a proposal to tax cyclists pitted cyclists' needs against country-based interests intent on building country roads. The 1930s debate was a replay of the 1890s. This time—like in the Netherlands—the coalition did not manage to halt the tax.[47] The Chamber of Commerce and Labor as well the Chamber of Industry and Trade opposed, but the Chamber of Agriculture saw the tax as an opportunity to finance country-based roads. In coalition with the cyclists' federation, local gun and bicycle manufacturer Steyr-Daimler-Puch AG petitioned the parliament of Styria. If the tax passed, the petition argued, the company would lose business and workers their jobs. As the "cheapest transportation," bicycles were the shoes for the employed and unemployed: their taxation therefore was "outrageous" as well as "regressive and asocial."[48] After the longest debate in Styria's parliamentary history, the controversial bicycle tax proposal nevertheless became law. It did not end there. A young communist, Josef Duschill, organized protests together with the party's youth organization (*Jugendverband* KJV).[49] There were petition drives in Graz and Knittelfeld. Together with the Cyclist Gauverband and the Trade Union Federation's Sports Department, a 36-year-old assistant at Graz University of Technology, Ing Traugott Schiffmann, collected six thousand signatures for a petition against the tax. The Nazi security troops, however, found the procedure "questionable" and forbade him to continue.[50]

What the communists and Schiffmann failed to accomplish, the Right did. The Austrian *Anschluss* regime soon relented and abolished the controversial tax in 1938. Well attuned to the grassroots resentment, right-wing parties saw the bicycle tax controversy as an opportunity to preempt the Left in competing for the working-class vote. Austria's right-wing political success was gratefully copied elsewhere. The Fascists in Italy did the same in 1939. In 1941 soon after they occupied the Netherlands, German authorities also abolished the despised bicycle tax. The repeal fitted seamlessly with the overall Nazi policy of currying political favor with the working classes to break the resistance of the Left.[51] It was foremost an opportunistic move; otherwise, the Nazis implemented a host of pro-car policies and bicycle-unfriendly measures. They introduced the engineering design principle of distinguishing fast and slow traffic; prioritized fast traffic by abolishing the rule that cyclists coming from the right "had priority"; and most notably as part of the military build-up, perfected the autobahn system that excluded what was then called "slow traffic": cyclists, pedestrians, and carts.[52]

The abolishment of the bicycle tax confronted the state how crucial the revenues had been for the Dutch national road-building effort. Where it had previously contributed 15 percent, the government was forced to pick up over 50 percent of the bill during the war when the bicycle tax no longer flowed into the fund and revenues from motorists failed to make up the difference.[53] By the Second World War, however, state funding was no longer controversial. By then, (fascist, welfare, and liberal) states gladly subsidized road building for motorized traffic; it was cast as a national priority embedded in a new definition of public good.

Historian Anne-Katrin Ebert has argued that the introduction of the bicycle tax in the 1930s gave cyclists political leverage to demand the construction and maintenance of cycling infrastructure. As we have seen, the Dutch bicycle tax not only disproportionally contributed to the road funds that financed the construction of new roads for motorized traffic, but the paths were also designed to discipline rather than facilitate cyclists, expressing the increasingly dominant—and internationally supported—view that separating motorists and cyclists was the only way to ensure safety and speed for a future dominated by cars. Most importantly, these rural roads did not benefit the taxpaying urban crowd.[54] That said, with the repeal of the bicycle tax, the government's already extremely minimal investments in bicycle paths along national and provincial highways and other infrastructures did disappear almost entirely; (middle-class) cyclists thus lost even the very modest leverage that they had possessed previously. The few extant bicycle lanes were soon eyed by authorities as convenient parking spaces for cars.[55] For the next four decades, cyclists disappeared from the political agenda as a legitimate voice in the discussion about the public good, despite the exploding numbers of cyclists during and after the war.

Sustainability and Tax Debates in Europe and the United States since the 1970s

After almost a four-decade hiatus, the Dutch national government once again began to allocate modest funds for bicycle infrastructures, this time in response to a powerful grassroots movement.[56] The reframing of the city bicycle as ecofriendly came from a grassroots movement that spread transnationally to many cities.[57] By reframing it thus, activists succeeded in moving the city bicycle from the political margins (as individual, working-class transportation) to the center (as the vehicle of sustainable mobility).[58] Bicycle activism was also an urban rebellion against the nation-state and its rural support base. In this third period of the bicycle debate that started in the 1970s, urban elites redefined the utility bicycle as key to creating more livable and sustainable cities. Through this reframing, the movement succeeded in shedding the bicycle's negative working-class associations, away from De Sica's proletarian protagonist. The discursive shift to the green citizen represented a major political realignment of class.

In the early 1960s, bicycles acquired the image of sustainable mode of transport when the Provo movement in Amsterdam placed them at the heart of its broad-based, antigrowth, and countercultural movement. In 1966, Provo, with sympathy for the working-class underdog, but with no links to organized labor or traditional labor politics, proposed a community-based bicycle scheme. Everybody would share white painted bicycles for free, as a common public good accessible to all.[59] Their bicycle-based vision came at the very moment that many traditional labor politicians (from Sweden in the 1950s to the Netherlands in the 1960s) began to support the auto industry as the basis for full employment and had argued that all working-class families should be able to afford a car in the near future.[60]

The anarchistic Provo ideas echoed throughout the emerging countercultural green movement. By the mid-1970s, the newly established national Only True Bicycle Union (ENFB) engaged in a political confrontation with the former cycling club and tourist organization ANWB, accusing it of having sold out to the automobile lobby. Together with its international counterparts in Rennes, Lyon, Paris, Graz, Vienna, Copenhagen, and Portland, the organization promoted the bicycle as a form of environmentally sustainable mobility.[61] It played a crucial role in challenging assumptions about automobility and the public good. No longer did the debates focus on filling the national treasury and building car infrastructure on the premise that cycling was a dying art. To the surprise of policy makers, many had continued to cycle. Debates now focused on encouraging cycling for both better mobility and a healthier environment. Tax incentives for individual commuting cyclists were introduced as one policy instrument; introducing public-private bicycle sharing programs another. They represented two visions of the public good. One assigned a central role to the national government; the other placed faith in public-private partnerships between local communities and multinational corporations as a way of avoiding the politically more sensitive issue of directly raising taxes for a more sustainable future.

Between 1965 and 1983, bicycle activists succeeded in advancing their ideas and expertise throughout the European and American urban world. Policy makers in cities faced congestion and citizens upset by the destruction of streets for faster automobility (a situation that the transnational community of engineers in national state agencies had helped to create). By the late 1980s, urban politicians and policy makers gave in to the pressure of civil-society organizations and agreed that motorized traffic would only continue to increase. Sooner or later, cars would completely clog the arteries and heart of the city beyond repair. Several countries developed urban and national mobility policies. Cities from Strasbourg in France to Erlangen in Germany introduced bicycle master plans.[62] French policy makers shared expertise and best practices nationally through the Club des Villes Cyclables, established in 1988. In the Netherlands, the grassroots cycling union ENFB took the lead in pushing fiscal measures to promote cycling as socially and environmentally beneficial through workshops, publications, information, and demonstrations.[63] Dutch policy makers responded to such national grassroots activism with new bicycle tax policies that served as a model worldwide.

In the late 1980s, the Dutch Ministry of Transport under Nelie Smit-Kroes made a small step in discouraging commuting motorists. The ministry's research showed that most (suburban) motorists used their cars for only short distances that bicycles could easily bridge. At the urging of the grassroots bicycle movement and with parliamentary support, she proposed in 1989 to cap the commuter credit for automobiles at 30 kilometers.[64] It was a modest reversal of the 1950s pro-car tax policies. Two years later, the minister issued a policy to reduce car traffic by 30 percent in 2010. The policy sought to encourage public transportation and cycling while attempting to raise car taxes—a debate that is still ongoing and unsolved.[65] Written in collaboration with the Dutch cycling union, the Ministry's 1990 Bicycle Master Plan invested approximately €20 million in constructing bicycle infrastructure and increasing cyclists' safety. A pilot program encouraged commuter bicycles (*bedrijfsfietsenplan*).[66]

The tax office, however, was not in line with the ministry's new mobility policies: cycling employees were taxed higher than employees who commuted by car. The cycling union protested the inequity.[67] Social democratic Willem Vermeend garnered broad parliamentary support in protesting the unequal taxing of the "sturdy weather-proof Dutch commuter bicycle."[68] Public pressure for bolder measures increased. From the sensationalist to the conservative protestant, newspapers ridiculed the government's timid proposals for the public at large while the ministry was experimenting with a bicycle lending program for its own employees.[69] More problematic was that the business community found parts of the taxing scheme too opaque to bother. In 1993, the ministry established a working group to clear barriers for the severely underused tax incentive.[70]

Civil society groups like Cycling Civil Servants Naaldwijk (FAN) and the Dutch Cycling Union were crucial for the ultimate success of the new bicycle tax incentive. Together they implemented a host of measures bicycles leases, loans, donation programs, and other financial compensation that the Dutch tax agency found too difficult to handle. FAN began to act as the national spokesperson for all municipal officials and soon merged with the *Foundation Cycling to work? Do!*—a nonprofit organization sponsored by the government and the bicycle industry.[71] Pressured by the cycling union, the Foundation, and the bicycle industry, the new labor undersecretary, Vermeend, successfully introduced a bicycle commuter scheme in 1994.[72] Employers could give their workers a bicycle every three years with a value up to 1500 guilders (in 2014 up to €749) for commuting under 15 kilometers through an income tax credit of 150 guilders (€100). Employers also could deduct from their corporate taxes for offering their employees free raingear, bicycle locks, maintenance services, and panniers worth up to 550 guilders (€227 today's rate). In collaboration with the Foundation, the government invested in the Bicycle Information Point (FIP) to help employees, unions, and employers navigate the new tax incentive. The Foundation expanded its mandate beyond the tax and promoted bicycle facilities like bicycle lanes, parking, and transfer areas at city edges.[73] These bottom-up initiatives helped turn the government's company bicycle program into an overwhelming success. In 2008 alone, almost one-fifth of all new bicycles (240,000 total, averaging €836 each) were financed under the bicycle commuter program.[74]

Various European governments soon adopted the Dutch tax incentive policy. Belgium introduced the new commuter tax incentives in 1997; research showed that companies who made use of the law saw the commuter cycling of their employees increase by 50 percent (from 6.3 to 9.5 percent). The British government's 1999 Green Transport Plan encouraged employers to lend their workers bicycles and bicycle clothes for a small monthly fee. By 2005, it had transformed into the Cycle to Work tax incentive scheme, enabling employers to lend their employees tax-free commuter bicycles. In the United States, riding the wave of President Obama's electoral victory, activists and the bicycle industry succeeded in passing the Bicycle Commuter Act through US Congress in 2008 as part of the legislation to promote renewable energy; the law encouraged employers to build bicycle parking facilities, showers for cyclists, and bike paths. Not all initiatives were successful. The Italian government issued a 30 percent refund program on all new (electric) bicycles that had so many limitations that it had little effect.[75] In France, the urban bicycle policy makers of the *Club des villes & territoires cyclables* failed to get similar measures passed.

There have been other indications that the push for tax equity between cyclists and motorists was rather modest. In 2002, the European Commission presented a new strategy to remove tax obstacles impeding the free movement of passenger cars within the EU's internal market. In 2009, the European bicycle industry demanded the same treatment for bicycles. The Comité de Liaison des Fabricants Européens de Bicyclette (COLIBI), the Association of the European Two-Wheeler Parts & Accessories' Industry (COLIPED), and the European Two-Wheel Retailers' Association (ETRA) jointly appealed to EU Taxation Commissioner László Kovács for a European strategy harmonizing tax incentives for cycling and lobbied to lower sales taxes on all bicycle products and services. The measure succeeded in some countries, but failed in others. In fact, the European Union has continued to focus most policy attention on encouraging motorized mobility.[76]

The European lobbying efforts for tax equity point to the inconvenient truth. The Dutch-inspired international tax incentives for individual commuting cyclists might have been an impressive national policy instrument, but the measure never fundamentally questioned automobility or reallocated funds on the basis of costs-and-benefit analysis for sustainability. The hardening of the automobility focus both nationally and through the European Union has been reinforced by the shifting governance mandate from national to city governments. Since the 1980s, as neoliberalism gained steam, cities have been facing retreating national governments. They have lost both state funding and the (modest) bicycle mobility expertise that national agencies had developed.

To be sure, cities gained more local freedom to manage their own affairs, creating a new political niche for mobility solutions for the urban constituencies that national governments had disregarded and suppressed for so long. It represented a fundamental redefinition of what constituted the public good in terms of urban mobility needs. In 1976, inspired by the Dutch Provo movement, the environmentalist Mayor Michel Crepeau in La Rochelle successfully instituted the first bicycle share program with three hundred self-service bicycles. Subsidized by a user-fee system and local and national funds, this was a publicly funded mobility plan. From Paris to Barcelona, many socialist and environmentalist mayors followed Crepeau's example to promote cycling policies. In the 1990s, urban politicians used bicycle-share schemes as a tool for city branding and solving congestion. They were often implemented in response to local electoral politics.[77]

Yet, these successes mask another inconvenient truth. Most urban politicians introduced public-private bicycle-sharing programs at relatively low costs to deal with clogged streets and meet environmentalists' demands.[78] Because no national governmental funds or support were available, as cycling had been removed from national definitions of the public good, cash-strapped cities everywhere jumped on the opportunity to outsource the costs of urban street furniture to multinational corporations. The new business scheme for funding cities' public mobility began in Lyon in 1965. As Maxime Huré has argued, Lyon's partnership with the French advertising agency JCDecaux would become a political alternative to raising taxes for urban services. In exchange for installing and maintaining the street furniture that cities found hard to fund, JCDecaux got a contract for outdoor advertising, expounding the appeal of public transport sites to generate revenues. The company turned the Lyon experience into a multibillion-euro and transnational business by sponsoring street furniture from benches to bus stop shelters.[79]

Bicycle sharing became the corporate leverage to break open a lucrative market of public space. To break JCDecaux's near monopoly in Europe's outdoor advertising, the US communication and outdoor advertising company Clear Channel joined in the urban bicycle-sharing trend in 1998. The American company offered to fund an innovative, free bicycle-share scheme with two hundred bicycles and twenty-five stations in the French city of Rennes. The program put the provincial town on the international map of city branding for its smart mobility expertise. Funding bicycle schemes, however, offered the multinational corporations a foot in the door and political leverage in the fierce transnational competition for outdoor advertising space. Subsequently, the French multinational JCDecaux—following Clear Channel's example—won the coveted Paris ten-year contract, spending about $140 million and employing 300 full-time staff to operate and maintain 20,000 bicycles in 1800 stations. The city received all revenues from the bicycle program, but the corporation's revenues from its 1628 city-owned billboards are a well-kept secret.[80] And that is indeed the second inconvenient truth.

These public-private contracts, while seemingly free for the public at large, made cities depend on multinational advertising companies like JCDecaux, Clear Channel, and the German Wall. Some cities have sought to navigate the political perils of becoming too dependent on corporate money. Under Catalan socialist and environmentalist mayor Jordi Hereu, Barcelona also partnered with Clear Channel for its bicycle-share scheme. The city managed to avoid complete dependence because early on, the mayor had secured the car park income as revenue to support the bicycle infrastructure. Barcelona's *Bici* system, however, was the exception in shifting funding from car to cycling infrastructure.[81]

Most bicycle initiatives have avoided the sensitive issue of raising or reallocating car-based revenues for more sustainable cities. And in all these schemes, the national governments and public works agencies that had so lavishly funded car-governed infrastructures have been mute on the subject. Cities and the urban multitudes are left to fend for themselves once again. The nonprofit Bicycle Transportation Alliance (BTA) in Portland, Oregon, has sought to address that issue head on. In its support of the Oregon state government's proposal to introduce a bicycle tax, the alliance argued that bicycle taxation would increase cyclists' profile and influence. Formed in 1990, the BTA considered bicycle taxes a tool of political leverage to demand the development of cycling infrastructure and create a viable bicycle lobby.[82] Thus the bicycle manufacturers and activists became allies in actively promoting bicycle use, much as they had in the 1890s. This time they mobilized the discourse on health, safety, sustainability, environment, and traffic congestion through the instrument of taxation. Today's tax incentives aim to encourage rather than curtail cyclists.

Yet, the fundamental class bias against urban cycling has not been solved, nor the willingness to allocate national funds to pay for sustainable mobility. A city forum in the bicycle city Münster in Germany organized an Internet referendum in 2011 on whether to introduce an annual bicycle tax of €5 to fund bicycle-related measures.[83] Taxes would help to reduce cycle theft, discipline cyclists, create bicycle-related employment, and support traffic flow (*Hilfreich im Strassenverkehr*).[84] In 2012, citizens in Bonn bluntly questioned why cyclists should be exempt from paying taxes with the result that they contribute nothing to the road system while motorists' taxes are wasted on bicycle lanes that cyclists ignore anyway. A year later, a CDU councilor in Leipzig called for a bicycle tax to cover the cost of snow removal and de-icing of bicycle lanes in response to what he believed were the outrageous demands of the General German Bicycle Club (ADFC). The organization had argued that if authorities failed to clear the bicycle lanes that cyclists were legally bound to use (*Benutzungspflicht*), then they should instead have full access to cleared (motor) roads.[85] In Stuttgart, bicycle tax proposers linked the measure to the widely shared sentiment that cyclists' offensive behavior should be disciplined: the tax registration system would conveniently enable police and citizens to identify unruly cyclists who knowingly violate traffic rules. On the Internet, all these proposals were overwhelmingly rejected.

In Switzerland and Liechtenstein, the bicycle tax was coupled with a registration and license plate system that forced cyclists to take out accident-liability insurance (*Unfall-Haftplichtversicherung*), relaying the underlying message that because cycling is an inherently risky business, any damage incurred by cyclists to others should be financially and legally covered by cyclists as a separate group of road users. Individual citizens should not be liable for such risk.[86] These discussions, however, implied cyclists were by definition dangerous road users, and should be punished for their irresponsible behavior with a bicycle tax and otherwise be liable.[87] In short, cyclists were not legitimate road users, whose mobility should be protected, facilitated, and advanced for the greater good.

Conclusion

As we have shown, in the first phase between the 1890s and 1920s, civil society actors saw the bicycle as a recreational and elitist mode of transport that signified an emancipated and good lifestyle. In the second phase, between the 1920s and 1940s, state actors and car lobby organizations successfully defined the bicycle as a utilitarian, slow, old fashioned, dangerous, and troublesome mode of transport for the working class. In the third phase, since the 1970s, public and private actors successfully redefined the bicycle as a fast, cheap, healthy, clean, sustainable, and morally superior mode of transport for green citizens. These cultural notions reappeared in public good and taxation debates that have hardened in traffic policies. The new images and meanings of the bicycle have not, however substituted existing ones. In fact, they coexist and even compete. Cycling activists have been able to reinstate bicycles as potential tools for sustainability since the 1970s, but old representations of cyclists as illegitimate road users die hard. We still hear the class-biased echoes of the past in identifying cyclists as undisciplined, irresponsible road users. We still hear old funding priorities. Although the Netherlands are often presented as a cycling paradise, its (tax and traffic) policies never fundamentally questioned automobility or reallocated funds. The bicycle was granted some space beside the dominant car in a social context of "peaceful coexistence."

There is a larger point for innovation studies as well. We see how innovations like bicycles can survive and compete. We also see how they reappear as a modern mode of transport, providing food for thought on how such innovations die out but also start (anew). Despite resistance, the "obsolete" and "working-class" bicycle has persisted since the 1970s. User-based organizations have played a crucial role in its reappearance. They have not only reused old technologies; they also have given them new meanings.[88] And cultural framings do make all the difference.[89] Cycling offered the most serious alternative to—and was the greatest competitor of—motorized mobility for more than six decades. In our common narratives, however, the bicycle simply lost out. Instead of the alleged inherent technological superiority of automobiles over bicycles, it was power politics, expert knowledge, and social class that won the fight. And yet, even this trajectory is too simple. In practice, bicycles and automobiles rode in tandem for at least seven decades; subsequently, the bicycle survived the forty-year dominance of the car before being reintroduced in recent decades as a more viable vehicle for a sustainable future. As historians of technology and innovation scholars, we are called to acknowledge the simultaneity of the old within the new by understanding how the cultural frame of sustainability radically reframed the cultural meaning of the bicycle. Yet in terms of the environmental challenges, the question, though, is whether the cultural framing will be reinforced by the politics of allocating the much-needed funds to make the transition toward sustainability possible.

Adri Albert de la Bruhèze is assistant professor of history of technology at the University of Twente, Faculty Behavioral, Management and Social Sciences (BMS), Department of Science, Technology and Policy Studies (STePS). He has published on the history of radioactive waste management in the United States, the history of bicycle use in Europe, food and nutrition history in The Netherlands, the emergence of Dutch consumer society, the history of technology in twentieth-century Netherlands, and transnational European tourism regimes. His most recent publications include the seventh volume of *Techniek in Nederland in de Twintigste Eeuw* (Technology in the Netherlands in the Twentieth Century, 1998–2003) with Johan Schot, Harry Lintsen, and Arie Rip; and *Manufacturing Technology, Manufacturing Consumers: The Making of Dutch Consumer Society* (2009) with Ruth Oldenziel.

Ruth Oldenziel is a professor at Eindhoven University of Technology and a fellow at the Rachel Carson Center, Munich, 2013–2015. She received her PhD in American history from Yale University (1992). Her publications include books and articles in the area of American, gender, and technology studies, among them *Hacking Europe*, edited with Gerard Alberts (2014); "Recycle and Reuse," edited with Heike Weber (special issue of *European Contemporary History* 2013); *Consumers, Users, Rebels* (2013) with Mikael Hard; *Cold War Kitchen*, edited with Karin Zachmann (2009); *Gender and Technology*, edited with Arwen Mohun and Nina Lerman (2003). She is currently working on a monograph, "Global Cycling: Paths towards Sustainability."

Acknowledgements:

The authors are grateful to Cindy Ott, the participants of the works-in-progress seminar at the Rachel Carson Center, LMU, Munich, the members of the "Long-Term Development" group, Department of Science, Technology and Policy Studies (STePS), Faculty Behavioral, Management, and Social Sciences, University of Twente, and the anonymous reviewers for their comments.

Notes

1. Ruth Oldenziel and Mikael Hård, *Consumers, Users, Rebels: The People Who Shaped Europe* (London, 2013), 160–161.
2. For a useful overview of the international discussion on bicycle taxation and registration, see "Why does the motorized majority want cyclists to be taxed and pay licenses & number plates," http://ipayroadtax.com/licensed-to-cycle/licensed-to-cycle/ (accessed 9 November 2013).
3. Scholarship on the bicycle tax and its meaning is lacking with the exception of James Longhurst, whose rich US-based case study focuses exclusively on the 1890s and introduced the notion of the public good to the discussion. Longhurst, "The Sidepath Not Taken: Bicycles, Taxes, and Rhetoric of the Pubic Good in the 1890s," *Journal of Policy History* 25, no. 4 (2013): 557–586; Longhurst, "'Awheel from Chicago to the Twin Cities': Legacies of Turn-of-the-Century Bicycle Paths in Minneapolis and St. Paul," in *Two Cities, One Hinterland: An Environmental History of the Twin Cities and Greater Minnesota*, ed. George Vrtis and Christopher W. Wells (2015). While not on taxes as such, see also Anne-Katrin Ebert, "When cycling gets political: Building cycling paths in Germany and the Netherlands, 1910–40," *Journal of Transport History* 33, no. 1 (2012): 115–137.
4. *Black's Law Dictionary*, 5th ed. (1979), 1307.
5. Ferdinand H. M. Grapperhaus, *Over de loden last van het koperen fietsplaatje. De Nederlandse rijwielbelasting 1924–1941* (Deventer, 2005), 72–73; A. B. Atkinson, "Optimal Taxation and the Direct versus Indirect Tax Controversy," *Canadian Journal of Economics* 10, no. 590 (1977), 592; http://www.investorguide.com/igu-article-1138-tax-basics-what-is-difference-between-direct-and-indirect-tax.html (last accessed October 2012). Taxes are levied either *directly* or *indirectly*. Direct taxes are based on the individual taxpayer's profile and include income taxes, social security contributions, and property taxes, whereas indirect taxes are imposed on transactions and activities irrespective of the circumstances or buyer/seller's profile, like sales taxes (VAT).
6. Robert Cavallès, "La taxe sur les vélocipèdes" (PhD thesis, Université de Toulouse, 1908), appendix. See also Richard Holt, "The Bicycle, the Bourgeoisie and the Discovery of Rural France," *British Journal of Sports History* 2 (1985): 127–139, 128.
7. Donald Weber, *De blijde intrede van de automobile in België, 1895–1940* (Gent, 2010), 22–23. France abolished the tax in 1958; and Belgium only during the period 1986–1991. Stefano Pivato, "The bicycle as a political symbol: Italy, 1885–1955," *International Journal of the History of Sport* 7, no. 2 (1990): 172–187; Catherine Bertho Lavenir, *La roue et le stylo. Commes nous sommes devenus touristes* (Paris, 1999).

8. Longhurst, "The Sidepath Not Taken," 561–562. Longhurst, "Awheel from Chicago to the Twin Cities," 5.
9. Ibid., 1–6.
10. Volker Briese, "From Cycling Lanes to Compulsory Bike Path: Bicycle Path Construction in Germany, 1897–1940," in *Proceedings of the 5th International Cycle History Conference* (1994), 124–126.
11. Ibid., 124–126; Frank C. A. Veraart, "Geschiedenis van de fiets in Nederland, 1870–1940" (MA thesis, TU Eindhoven, 1995).
12. Ibid., 33–35, 66–67; Anne-Katrin Ebert, *Radelnde Nationen. Die Geschichte des Fahrrads in Deutschland und den Niederlanden bis 1940* (Frankfurt, 2010), 378–390.
13. Adolf W. K. Hochenegg, "Radfahrsteuer oder nicht?" (Leipzig, 1898); Weber, *De blijde intrede van de automobile in België*. Cavallès, "La taxe sur les vélocipèdes." See also Oldenziel and Hård, *Consumers, Users, Rebels,* chap. 4.
14. Weber, *De blijde intrede van de automobile in België,* 20, 23.
15. William Everett Hicks, "Shall We Tax the Human Leg?" *North American Review* CLX (October 1897): 512; Cavallès, "La taxe sur les vélocipèdes"; Richard J. B. Bosworth, "The Touring Club Italiano and the Nationalization of the Italian Bourgeoisie," *European History Quarterly* 27, no. 3 (1997): 371–410, here 380; Bernd Kreuzer, "1 Fahrrad = 0,25 PKW-Einheiten: Das Fahrrad im Stadtverkehr zwischen verpaßten Chancen und gewollter Marginalisierung, Pfadabhängigkeiten und Gestaltungsspielräumen," in *Erfahrung der Moderne. Festschrift für Roman Sandgruber zum 60. Geburtstag,* ed. Michael Pammer, Herta Neiß, and Michael John (Stuttgart, 2007), 465–481, 274, 277.
16. Sue-Yen Tjong Tjin Tai, Frank Veraart, and Mila Davids, "How the Netherlands became a bicycle nation: Users, firms and intermediaries, 1860–1940," *Business History* (2014).
17. Bernd Kreuzer, "Das Fahrrad in Oberösterreich: drei Unternehmen, ein Nebenprodukt und seine Nutzung seit 1870," in *Technikland Oberösterreich. Wirtschaftliche Entwicklungen und industrielle Gegenwart; Syposium, Linz, 22. und 23 Jänner 2010,* ed. Ute Streitt (Linz, 2013), 151–166; Hochenegg, "Radfahrsteuer oder nicht?" 11–14; Martin F. Polaschek, "Funktionierender Parlamentarismus im Ständestaat? Die Auseinandersetzungen um die Einführung einer Fahrradabgabe in der Steiermark," *Zeitschrift des Historischen Vereines für Steiermark* LXXXVI (1995): 277–301, here 278–279. For France, see Cavallès, "La taxe sur les vélocipèdes."
18. Longhurst, "The Sidepath Not Taken," 564; Hochenegg, "Radfahrsteuer oder nicht?" 16.
19. Hochenegg, "Radfahrsteuer oder nicht?" For France, see Cavallès, "La taxe sur les vélocipèdes"; for Italy, Bosworth, "The Touring Club Italiano," 380.
20. Oldenziel and Hård, *Consumers, Users, Rebels,* chap. 4.
21. Longhurst, "The Sidepath Not Taken."
22. On Infrastructures, see Wolfram Kaiser and Johan W. Schot, *Writing the Rules for Europe: Experts, Cartels, International Organizations* (London, 2014); Per Högselius, Arne Kaijser, and Erik van der Vleuten, *Europe's Infrastructure Transition: Economy, War, Nature* (London, forthcoming).
23. Ruth Oldenziel and Adri Albert de la Bruhèze, "Contested Spaces: Bicycle Lanes in Urban Europe, 1900–1995," *Transfers* 1, no. 2 (2011): 31–49.
24. Ibid.; Oldenziel and Hård, *Consumers, Users, Rebels,* chap. 4.

25. Gijs Mom, "Decentering Highways: European National Road Network Planning from a Transnational Perspective," in *In der Moderne Strasse: Planung, Bau und Verkehr vom 18. bis zum 20. Jahrhundert,* ed. Hans-Liudger Dienel and Hans-Ulrich Schiedt (Frankfurt, 2010), 77–100, esp. 80, 94–90.
26. Ruth Oldenziel, "The Vanishing Trick: How Cyclists and Pedesterians were Left Uncounted" (forthcoming); Mom, "Decentering Highways," 90–93; Denis Pye, *Fellowship is Life: The National Clarion Cycling Club 1895-1995* (Bolton, 1995); David Prynn, "The Clarion Clubs, Rambling and the Holiday Associations in Britain since the 1890s," *Journal of Contemporary History* 11, no. 2/3 (1976): 65–77; Ruediger Rabenstein, *Radsport und Gesellschaft: ihre sozial-geschichtlichen Zusammenhaenge in der Zeit van 1867 bis 1914* (Hildesheim, 1955); Briese, "From Cycling Lanes to Compulsory Bike Path"; Aage Hoffmann, "Arbejderidr æ ttens forhold til socialdemokratiet ca. 1880–ca. 1925," *Arbejderhistorie* 1 (2008): 96–115; Pivato, "The bicycle as a political symbol."
27. Gijs Mom, "Decentering Highways," 87; Mom, "Building an Infrastructure for the Automobile System. PIARC and Road Safety (1908-1938)" (Paris, 2007); Adri Albert de la Bruhèze and Frank C. A. Veraart, *Fietsverkeer in praktijk en beleid in de twintigste eeuw. Overeenkomsten en verschillen in fietsgebruik in Amsterdam, Eindhoven, Enschede, Zuidoost Limburg, Antwerpen, Manchester, Kopenhagen, Hannover en Basel* (Eindhoven, 1999); Bernd Kreuzer, "Verkehrszählungen auf Österreichs Straßen zwischen den Weltkriegen und ihr Beitrag zur Vekehrsgeschichte," *Blätter für Technikgeschichte* 60 (1998): 43–62, here 50; Oldenziel, "The Vanishing Trick."
28. Italo Bonardi, "Ciclo-moto-turismo: Le bachine ciclistiche," *Le vie d'Italia* (January 1938): 9–13, as quoted in Carlos Héctor Caracciolo, "Bicicleta, circulación vial y espacio público en la Italia Fascista," *Historia Critica* 39 (2009): 20–42, here 30. See also Pivato, "The bicycle as a political symbol," 173.
29. Briese, "From Cycling Lanes to Compulsory Bike Path"; Caracciolo, "Bicicleta"; Oldenziel and Hård, *Consumers, Users, Rebels,* chap. 4.
30. Mom, "Decentering Highways," 78–100.
31. Ibid., 77–82.
32. Mom, "Building an Infrastructure for the Automobile System"; Oldenziel, "The Vanishing Trick."
33. Bruhèze and Veraart, *Fietsverkeer in praktijk en beleid,* 44–47.
34. Gijs P. A. Mom and Ruud Filarski, *Van transport naar mobiliteit. De mobiliteitsexplosie (1895-2005)* (Zutphen, 2008).
35. Peter E. Staal, *Automobilisme in Nederland: Een geschiedenis van gebruikers, misbruik en nut* (Zutphen, 2003), 73; Bruhèze and Veraart, *Fietsverkeer in praktijk en beleid*; Kreuzer, "Verkehrszählungen auf Österreichs Straßen."
36. Mom and Filarski, *Van transport naar mobiliteit,* 197.
37. Bruhèze and Veraart, *Fietsverkeer in praktijk en beleid,* 46.
38. Table based on S. A. Reitsma, "De rijwielbelasting als bestemmingsheffing voor het motorsnelverkeer," *De Opbouw* 2 (1938), 5, 10, 13, 18, 19, 20, 21; S. A. Reitsma, *Herwaardering van Verkeerseconomische Waarden* (Den Haag, 1942), 48, 54, 56, 57; CBS, *Statistisch jaarboek van de jaren 1927-1935* (Den Haag, 1936); CBS, *Verkeersstatistiek 1933-1944* (Den Haag, 1944); CBS, *Maandschrift* (January 1937); "Wet van den 30sten December 1926, tot het heffen van eene belasting en treffen verdere voorzieningen

ten behoeve van de openbare verkeerswegen te land, hoofdstuk 11, Slotbepalingen," *Staatsblad* 464; Matea F. A. Linders-Rooijendijk, *Gebaande wegen voor mobiliteit en vrijetijdsbesteding. De ANWB als vrijwillige organisatie 1883–1937* (Heeswijk-Dinther, 1989), 274; Veraart, "Geschiedenis van de fiets," 97; Bruhèze and Veraart, *Fietsverkeer in praktijk en beleid*, 46, 84; G. J. van der Broek, "Financing of Government Roads," in *Permanent International Association of Road Congresses (PIARC), VIth Congress— Second Section: Traffic and Road Administration: Fourth Question, Ways and Means of Financing Highways* (Washington, DC, 1930); Mom and Filarski, *Van transport naar mobiliteit*, 97, 140, 161, 163, 197.

39. R. Tijken, "The Segregation of the Various Classes of Traffic on the Highway," in *Permanent International Association of Road Congresses (PIARC), VIIth Congress—The Hague 1938. Second Section: User, Regulation and Administration, 4th question*, 9–10. This claim was later confirmed by Albert de la Bruhèze and Veraart, *Fietsverkeer in Praktijk en Beleid*, 181; see figure 4.1, this chapter.
40. Mom, "Building an Infrastructure for the Automobile System."
41. Ebert, *Radelnde Nationen*; Ebert, "When cycling gets political"; http://www.uni-muenster.de/NiederlandeNet/nl-wissen/freizeit/vertiefung/fahrrad.
42. Mom and Filarski, *Van transport naar mobiliteit*, 198–199, 361–369. Peter Owen Engelke, "Green City Origins: Democratic Resistance to the Auto-Oriented City in West-Germany, 1960–1990" (dissertation, Georgetown University, 2011), 114–160.
43. An average workers' family income in the 1920s came to 15 guilders weekly or 800 guilders max annually. Maarten Dijkstra, "Lieve Kaatje, waar is je rijwielplaatje?" http://www.fietsbelasting.nl/?page_id= 452 (accessed 15 October 2012). Considering many did not have permanent jobs and depended on seasonal labor and economic fluctuations, this would amount to 3–10 percent of a worker's annual income.
44. Cavallès, "La taxe sur les vélocipèdes," 52, 94.
45. Leslie Schwartz, "Fietsbelasting in Nederland, 1924–1941," *Genealogie* 12, no. 2 (2006): 51–52; Grapperhaus, *De loden last van het koperen fietsplaatje*; Van der Poel, *Rijwielbelasting en de daarvoor gebruikte merken*.
46. "Fietsplaatje was een plaag, 17 jaren lang," http://www.rijwielbelasting.nl/geschiedenis.html.
47. Kreuzer, "Das Fahrrad in Oberösterreich."
48. Polaschek, "Funktionierender Parlamentarismus im Ständestaat?"; Wolfgang Wehap, *Frisch, Radln, Steierisch: Eine Zeitreise durch die regionale Kulturgeschichte des Radfahrens* (Graz, 2005), 121–123, 146–150. In the metropolitan area of Vienna, even a businessman without much sympathy for socialists' criticism of the Nazi regime articulated the same sentiment: "[for] thousands of unemployed, it [the bicycle] was the only practical method of getting around to look for jobs ... [and] for apprentices on very low wages it was almost indispensable." Charles A. Gulick Jr, "Vienna Taxes since 1918," *Political Science Quarterly* 53, no. 4 (1938): 533–556, here 552.
49. "Nummerntafeln und Abgaben auf Fahrrädern in der Steiermark Exkurs," http://graz.radln.net/cms/beitrag/10827820/25359581/ "Fahrradkennzeichen" fn 11, as accessed http://graz.radln.net/cms/dokumente/10827950_25359419/5fd12039/wago57.pdf (accessed October 2012). For a biographical sketch of Joseph Duschill, see: http://ooe.kpoe.at/article.php/20080702090004216, accessed October 15, 2012.
50. Wehap, *Frisch, Radln, Steierisch*.

51. An incident in the province of Limburg suggests the tensions when the Dutch mounted police (Marechaussee) fired at one of the NSB Weather Department employees who had failed to show a taxation tag; Schwartz, "Fietsbelasting in Nederland, 1924–1941," 53.
52. It was also forbidden for two cyclists to ride side by side or for cyclists to transport a passenger on the rear wheel luggage carrier. Pete Jordan, *De Fietsrepubliek* (Amsterdam, 2013), 152 and 157. For Nazi policies in general, see Thomas Zeller, "Building and Rebuilding the Landscape of the Autobahn 1930–70," in *The World Beyond the Windshield: Roads and Landscapes in the United States and Europe*, ed. Christof Mauch and Thomas Zeller (Athens, OH, 2008), 125–142. See also Uwe Fraunholz, *Motorphobia: Anti-automobiler Protest in Kaiserreich und Weimarer Republik*, Kritische Studien zur Geschichtswissenschaft (Göttingen, 2002).
53. Mom and Filarski, *Van transport naar mobiliteit*, 238.
54. Reitsma, "De rijwielbelasting als bestemmingsheffing." As the Dutch maverick economic statistician Reitsma already calculated in 1938, the bicycle tax was disproportionally generated by the numerous working-class cyclists. At the same time, its revenues were used for roads that sidelined rather than facilitated cyclists because they were built to benefit motorists. He argued that with the 1935 Traffic Fund, the costs for cyclists proportionally worsened because the bicycle revenues were used to support all modes of transport and their infrastructures: the government subsidized the rail and tramways that operated with great losses. We will deal with Reitsma in a separate essay. *Tijdschrift voor Sociaal-economische geschiedenis* (forthcoming).
55. Bruhèze and Veraart, *Fietsverkeer in praktijk en beleid*.
56. Ibid.; Mom and Filarski, *Van transport naar mobiliteit*, 339.
57. Oldenziel and Hård, *Consumers, Users, Rebels*, chap. 7.
58. Oldenziel and Bruhèze, "Contested Spaces," 40–42.
59. Zachary Mooradian Furness, *One Less Car: Bicycling and the Politics of Automobility* (Philadelphia, PA, 2010), 55–59; Hans Righart, *De eindeloze jaren zestig. Geschiedenis van een generatieconflict* (Amsterdam, 1995), 238.
60. Martin Emanuel, "Constructing the Cyclist: Ideology and Representations in Urban Traffic Planning in Stockholm, 1930–1970," *Journal of Transport History* 33, no. 1 (2012): 69–70.
61. Maxime Huré, "Les réseaux transnationaux du vélo; Gouverner les politiques du vélo en ville. De l'utopie associative à la gestion par des grandes firmes urbaines (1965–2010)" (Université Lyon 2 Lumière, 2013); Furness, *One Less Car*, 57–58.
62. On Strasbourg, see Huré, "Les réseaux transnationaux du vélo"; on Erlangen, see Engelke, "Green City Origins," 251–267.
63. "Fiets en Fiscus," *Actieradius: vervoermanagement in de praktijk* 2, no.3 (1994): 13–15; Wim Koehler, "Gezondheid: Goed voor hart en dikke darm," *NRC Webpagina's*, 26 February 1998, http://retro.nrc.nl/Lab/Profiel/Fiets/gezondheid.html (accessed 11 November 2012).
64. "Fiscale Discriminatie," *Nederlands Dagblad*, 19 March, 1990; "PVDA en CDA: Maatregelen voor milieu vriendelijker vervoer," *Het Vrije Volk*, 22 December 1989.
65. Ministerie van Verkeer en Waterstaat, *Tweede Structuurschema Verkeer en Vervoer*, Den Haag: Ministerie van Verkeer en Waterstaat, 1988; Tweede Kamer der Staten Generaal, vergaderjaar 1988–1989, 20922, nos. 1–2.

66. Tweede Kamer der Staten Generaal, Vergaderjaar 1990–1991, 20922, no. 98, 1–8, "Brief van de Minister van Verkeer en Waterstaat aan de voorzitter van de Tweede kamer der Staten Generaal"; "Kamer steunt Fietsplan May," *Telegraaf,* 26 September 1991; Tweede Kamer der Staten Generaal, Vergaderjaar 1990–1991, 20922, no. 107
67. "Fiscaal voordeel," *Nederlands Dagblad,* 11 March 1992; "Rijwiel van de baas mogelijk belastingvrij," *Limburgsch Dagblad,* 21 May 1992; "Van Amelsvoort overweegt belastingvrije leenfiets," *Nederlands Dagblad,* 22 May 1992.
68. "Stevige Hollandse fiets inzet van kamerdebat" *Limburgsch Dagblad,* 12 November 1992; "Kamer op bres voor 'gewone stevige' fiets," *Leeuwarder Courant,* 12 November 1992; "Woon-werkfiets doelwit fiscus," *Leeuwarder Courant,* 12 November 1992.
69. "Fiets," *De Telegraaf,* 13 November 1992; *De Telegraaf,* 12 November 1992; "Aanbod voor leasefietsen bij Provincie," *Limburgsch Dagblad,* 10 December 1993.
70. J. W. Gosselink, "Vervoer en Fiscus. Blijft de fiets de fiscale verschoppeling?" *Actieradius: vervoermanagement in de Praktijk* 3, no. 5 (1995): 14–15, here 14.
71. Koehler, "Gezondheid." There were other public-private initiatives: bicycle manufacturer Gazelle partnered with the tourist organization ANWB and the city of The Hague in a lease-bicycle program for their employees. "Woon-werkfiets doelwit van fiscus," *Leeuwarder Courant,* 12 November 1992.
72. Gosselink, "Vervoer en Fiscus," 15; E. Smit, "Als ik zoek ben, ben ik fietsen," *Fiets* 5 (2000) 44–9, here 47.
73. M. Dekker, "Hot item: fietsen met de fiscus—Ook een fiets van de zaak?" *Management Support Magazine* 14, no. 6 (1996): 8–9; "Fiscaal beleid ter simulering fietsgebruik," *PS: Periodiek voor Sociale Verzekering, Sociale Voorzieningen en Arbeidsrecht* 20 (6 September 1995): 1267–68; "Loonbelasting. Inkomstenbelasting. Omzetbelasting: Fiscale stimuleringsmaatregelen voor het gebruik van de fiets in het woon-werkverkeer," *Vakstudie-Nieuws: documentatie op het gebied van fiscaal recht* 51, no. 8 (1996): 669–673; W. Hulsman, "Douche voor Fietsende ambtenaren," *Reformatorisch Dagblad,* 25 March 2000, 43.
74. Colibi/Coliped/Etra Press Release, 24 April 2009, http://www.etra-eu.com/page}.asp?id=9303859 and http://www.cycling-embassy.dk/2010/07/12/tax-incentives-for-bike-commuting/ (accessed 13 July 2013). The tax incentive ended January 2014.
75. Ibid.; http://www.villes-cyclables.org/modules/kameleon/upload/cvtc_CP080909.pdf.
76. Five member states have implemented a reduced rate on bicycle repairs; Belgium, Luxemburg and the Netherlands apply 6 percent VAT, Poland 7 percent, and Greece 9 percent. Colibi/Coliped/Etra Press release, 24 April 2009.
77. Huré, "Les réseaux transnationaux du vélo."
78. Ibid.; Engelke, "Green City Origins."
79. Huré, "Les réseaux transnationaux du vélo."
80. Ibid. The exact amount is hard to come by. See also John Ward Anderson, "Paris embraces plan to become city of bikes," *Washington Post,* 24 March 2007.
81. Huré, "Les réseaux transnationaux du vélo," chap. 6.
82. In Colorado Springs, such dedicated tax has been a practice for twenty years. The $4 bicycle tax is allocated for constructing cycling infrastructure.
83. Many German cities have a digital citizen ("*Bürgerhaushalt*") service to map citizens' commitment and mood.

84. http://buergerhaushalt.stadt-muenster.de/ergebnisse-der-vorjahre/buergerhaushalt-2011/.
85. The proposal failed, partly because the Leipzig Green Party backed the cyclists' vision of mixed traffic; moreover everyone argued that Kempner's proposal made no legal sense because the German tax system precludes excise laws earmarked for one particular policy (*zweckgebundenen Steuern*); Matthias Puppe, "Fahrradsteuer in Leipzig: CDU Stadtrat relativiert Vorschlag—Abstimmung bei Facebook," LVZ-Online, 07.02.2012, http://www.lvz-online.de/leipzig/citynews/fahrradsteuer-in-leipzig-cdu-stadtrat-relativie.
86. Cees van Dam and Gerrit van Maanen, "The development of traffic liability in the Netherlands," in *The Development of Traffic Liability*, ed. Wolfgang Ernst (Cambridge, 2010), 112–150; Iris Borowy, "Road Traffic Injuries: Social Change and Development," *Medical History* 57, no. 1 (2013): 108–138.
87. Bürgerhaushalt Stuttgart, "Fahrradsteuer und Kennzeichen Einführen," https://www.buergerhaushalt-stuttgart.de/vorschlag/2911 (accessed June 2013).
88. Elizabeth Shove, "The Shadowy Side of Innovation: Unmaking and Sustainability," *Technology Analysis & Strategic Management. Innovation, Consumption, and Environmental Sustainability* 24, no. 4 (2012): 363–375; David Edgerton, *The Shock of the Old: Technology and Global History since 1900* (London, 2006).
89. Trevor J. Pinch and Wiebe E. Bijker, "The Social Construction of Facts and Artifacts: Or How the Sociology of Science and the Sociology of Technology Might Benefit Each Other," in *The Social Construction of Technological Systems*, ed. Wiebe E. Bijker, Thomas P. Hughes, and Trevor E. Pinch (Cambridge, MA, 1987), 17–50; Bijker, *Of Bicycles, Bakelites, and Bulbs. Toward a Theory of Sociotechnical Change* (Cambridge, MA, 1995);. Shove, "The Shadowy Side of Innovation," 368.

 CHAPTER 5

Monuments of Unsustainability
Planning, Path Dependence, and Cycling in Stockholm

Martin Emanuel

"When it [Centralbron, a bridge in central Stockholm] was planned around 1960, there was only one person (an engineer) who argued that bicycle lanes should be built on the bridge. To my knowledge he was not supported by a single politician. No bicycle lanes were thus built on Centralbron. Now the lack of bicycle lanes in this section is one of our difficult and insoluble problems. We are stuck. There is no room for bicycles on Centralbron."—Samuel Strandberg, 1980[1]

At a Nordic conference about vulnerable road users held in 1980, Samuel Strandberg, Liberal vice chairman of the traffic board in Stockholm, lamented the lack of bicycle lanes on Centralbron, one of very few connections between the northern and southern parts of the Swedish capital. During the 1960s, engineers and laymen had been caught up in a "car mania" that was reflected in their planning decisions. By 1980, it had been replaced by a more "sober" attitude to urban automobility (to use Strandberg's description). However, it turned out that it was not easy to undo past decisions, for they were literally cast in concrete.[2]

Stockholm is the "city on water," the "Queen of Lake Mälaren," the "Venice of the North." Homages to Stockholm never fail to praise the city's closeness to water and the views from its bridges. Any encroachment of the waterfronts and views of the lakes are closely followed and often fiercely debated from an aesthetic point of view. Once in place, however, the bridges have become symbols and important points of reference.[3]

At the same time, the city's location—it extends over several islands between the Baltic Sea and Lake Mälaren—also presents substantial challenges to maintaining good connections between its various parts and ensuring that it continues to grow and function effectively. In particular, connections between the north and south of Stockholm have been a major headache for planners and politicians. Their decisions tell us much about the technical challenges, the limitations, and the ideals of urban aesthetics, as well as economic realities and priorities.

Bicycles are currently understood as the ideal vehicle for more sustainable practices: they have found new purposes for new (policy) goals.[4] Elisabeth Shove stresses the particularities of reintroducing "old" technologies and practices into existing sociotechnical systems. After all, the disappearance of cycling in the postwar period was in most cases only relative and, importantly, a continuous process of reconfiguration between cycling and driving—in which cycling gradually moved from mass practice to a practice for a few, and from a "normal" practice to one not so normal—rather than a sudden substitution.

If cycling is to reemerge as an ordinary way of getting around, it will do so in an environment that is, in important ways, marked by the material and cultural traces of previous configurations of cycling and driving. On the one hand, Shove acknowledges the *durability* of past material and social cycling cultures; their capacity to survive despite, beyond, and alongside regimes and sociotechnical systems that have rendered them at least partially redundant. It is therefore crucial to analyze the continuity of fragments of regimes and social practices for understanding the conditions and chances of resurrection and revival of cycling over the last decades in particular.[5]

On the other hand—and this will be at the center of this chapter—the *obduracy* of existing (car-centered) urban transport systems, in which cycling is to reemerge, also needs to be considered. In spite of new purposes and policy goals, cultural as well as material remnants continue to hold cycling back. The infrastructures of urban transport are not neutral, but carry important power dimensions. They have been shaped by social actors in processes where cultural values, social relations, politics, and power are closely interwoven.[6] The planning of infrastructures is—just like policy instruments such as the bicycle tax treated in the previous chapter—political. It belongs to political scientist Steven Lukes's "third dimension" of power, that is, power exertion of a more subtle character than direct intervention: by changing the rules of the game, problem formulations, or what is considered self-evident and natural. For example, by designing infrastructure that naturalizes certain ways of moving around the city while deterring other ways.[7]

Infrastructures are thus the material remnants of past ideologies; they perpetuate past priorities and carry them into the future. The path dependency—defined by James Mahoney as the "historical sequences in which contingent events set into motion institutional patterns or event chains that have deterministic properties"—of infrastructures such as roads and rails is perhaps uniquely strong.[8] By continuing to shape our travel practices, these infrastructures are monuments of unsustainability, and the processes of un-building them are often complicated enough.[9]

In tracking such path dependencies, this chapter devotes special attention to the planning, construction, and partial reconstruction of two key infrastructures for urban mobility in Stockholm during the second half of the twentieth century: Centralbron and Essingeleden. Considering their importance for metropolitan traffic, the north–south connections can be taken as a fairly certain indication of shifts in the composition of and the visions surrounding urban traffic. The reconstruction processes provide an opportunity for understanding the negotiations between past and current configurations of cycling and driving.

Postwar Marginalization of Cyclists

In 1949, the staggering bicycle usage during the Second World War, which reached more than 70 percent at its peak, had dropped again to prewar levels of around 35 percent of all traffic. That same year, car traffic reached the very same share, having recuperated from wartime levels of around 10 percent. The following decades were a period of incremental but rapid conversion from cycling to driving. In 1950, 1960, and 1970, bicycle traffic decreased from 30 percent, to 3 percent, and then to less than 1 percent, respectively, of the total amount of traffic—to be compared with car traffic levels of 40, 75, and almost 90 percent.[10]

The postwar decline of cycling in Stockholm is all the more spectacular given the increase during the interwar period. The modal share of bicycle traffic increased (the only mode of transportation to do so) from 20 to over 30 percent during the 1930s.[11] In fact, the interwar period—not the postwar period—ought to be the period when the contestation between proponents of bicycles and cars was at its peak. Car ownership in Sweden had surged in the 1920s, and in urban and traffic planning circles the car was frequently called upon as the vehicle of the future. At the same time, the bicycle earned a reputation as a working-class vehicle, helping to suburbanize the masses. As a result of the high levels of cycling before, during, and just after the Second World War, cyclists were taken into consideration by the police, municipal engineers, and the bicycle lobby, primarily using short-term measures aimed at a more effective use of the existing street network as well as educating road users in this "effective use."[12]

Shove stresses the importance of the social status of cyclists vis-à-vis noncyclists for the configuration of cycling and driving.[13] Whereas the modal share of bicycle and car traffic in Stockholm was equal in 1949, the social position of cycling and driving was already radically different and moreover rapidly changing. The image of the bicycle was transforming from a working-class vehicle into a "poor man's vehicle"—allegedly used only by those who could afford nothing else.[14]

In Sweden, the trend was more dramatic than elsewhere in Western Europe.[15] The yearning for private automobility and "modern" public transport alternatives should not be disregarded as an important factor in the postwar decline of bicycle traffic in Stockholm. Other European countries had to spend their limited resources on rebuilding their war-torn economies and infrastructures, choosing a policy of keeping down consumption well into the postwar period. Politically neutral Sweden not only came out of the Second World War unscathed, but actually profited from the war; with a relatively sound economy, it had a much smoother peacetime conversion to consumerism. When the Swedish government eliminated all restrictions on private automobility that had existed prior to 1950, there was an overwhelming and rapid adoption of the car.[16]

These "pull" factors of consumption need to be balanced against an important "push" factor: the planning of urban traffic. In the booming economy of the postwar period, the position of urban and traffic planners was becoming increasingly important, and they seized the initiative in urban traffic matters. Their plans took a long-term perspective and included more extensive intervention into the built environment in order to cope with the increased "demands" of urban traffic. The bicycle (now cast as old-fashioned, irrational, and problematic) was completely absent from their future visions. Their interpretations of the bicycle as an unsafe, local, and primarily recreational mode of transportation were materialized in the urban infrastructure, which led to worse conditions for cyclists and reinforced the conversion to other modes of transport.

In the context of modernist planning, urban planners and policy makers did not regard the bicycle as a utilitarian mode of mobility; planners did not deem it capable of covering the commuting distances of what they now believed should be the standard: the functionally separated city. Thus, while carving out a small sphere for bicycle traffic in the suburbs, they made longer bicycle journeys to and from the city center more difficult. In the city center, policy makers chose a "do-nothing" strategy in response to the anticipated difficulties for cyclists. Admittedly, while they paid attention to the safety of cyclists in the suburbs, their measures often led to less comfortable conditions for cyclists—the safety focus effectively blocked other planning objectives, such as comfort and accessibility—and at times (as we will see below) the costs involved to ensure their safety were considered prohibitive.

Throughout this process, the capabilities of the bicycle were judged primarily in comparison with the dominant other—the car—with its very different needs in terms of vulnerability, speed, distance, and loading capacity. Thus, postwar urban transformation centered around the car and its needs (and in

Stockholm, the subway). Cyclists, on the other hand, were forced to cope with these changing conditions as well as they could.

Excluding Cyclists from Centralbron and Essingeleden

In the postwar period, then, urban planners and traffic engineers expelled Stockholm cyclists from the city using technocratic and seemingly neutral terms like *safety* and *capacity* to depoliticize what were, in fact, political decisions favoring the car.[17] Their exclusion of cyclists from Centralbron and Essingeleden are cases in point.

Centralbron, a bridge between the northern and southern parts of central Stockholm, was constructed in the late 1950s. Essingeleden was built slightly later as the western connection of a proposed ring road around the central region of Stockholm. Both were major infrastructure projects and important parts of the main road network. Both were, moreover, initially planned with bicycle lanes, which were however scrapped later on in the planning process—altering radically (given the importance of the north–south connection) and in a very material sense the possibilities for cycling (in relation to driving) the city. The postwar ideological reconfiguration of cycling and driving was hardened into infrastructure.

The building of Centralbron was approved by the city council in 1947, but in 1950 the project came up for review. The planned bicycle lanes on the bridge were reconsidered; omitting them would allow the road to become a through route for motor traffic. The Stockholm Central Board of Administration, supported by the head of the urban planning department's traffic bureau, Carl-Henrik af Klercker, was in favor of this omission.[18]

Prior to this, certain other important changes had been made to the proposed road: some of the midpoint entrance and exit ramps (the bridge passed by a smaller island in the historical center of Stockholm) had been taken away, turning the bridge into a part of a through traffic route rather than one for local journeys. According to af Klercker, this made the bridge a less appropriate choice for cyclists since, he argued, they wanted to have access to the local street network but were not interested in traveling "as quickly as possible along the complete length of the road section." Because bicycle use would be low, he argued that bicycle lanes were useless and the space could be better used for an extra car lane.

Another reason for scrapping the planned bicycle lanes was that the configuration of entrance and exit ramps at the end points of the bridge would be technically difficult if safe bicycle lanes were to be constructed. Safety considerations, in other words, only confirmed the appropriateness of scrapping the bicycle lanes and excluding bicyclists. Furthermore, bicycle traffic was de-

106 *Monuments of Unsustainability*

Figure 5.1. The planning document "Trafikledsplan för Stockholm" from 1960 outlined the future network of exclusive motor traffic routes in Stockholm. The two examples treated in this chapter, Centralbron and Essingeleden, were both included in the plan: Centralbron as part of the north–south connection in the very middle of the red circle (a ring road around the inner city of Stockholm), while Essingeleden is the western section of the ring road.

Source: "Trafikledsplan för Stockholm" (Stockholm: Stockholms stads generalplaneberedning, 1960).

creasing, and forecasts indicated that this trend would continue due to the "increasingly troublesome" conditions for bicycles as a result of increasing car traffic.

The planners were clearly not interested in preventing cyclists from being pushed aside by cars. Formulations to "relieve" stretches from bicycle traffic and create a traffic route purely for motorized traffic show that removing the bicycle lanes was planned to benefit car traffic. Before the final decision was made, the project was referred back to the Cyclists' Federation, which *supported* the proposal to scrap the bicycle lanes.

The case was really delicate, as af Klercker was a board member of the Cyclists' Federation from 1951 to 1953. The organization's main motive for approving the scrapping of the bicycle lanes was that the authorities had informally promised to build bicycle lanes on other streets. In its more comprehensive explanations, the Cyclists' Federation, in fact, echoed formal policy plans. By approving the removal of the bicycle lanes, the organization was clearly influenced by af Klercker, whose acceptance of automobility is evident in his writings.[19] Af Klercker considered automobility "unstoppable" and held that "urban planning must … after all be adjusted to the future human's ever more four-wheeled existence."[20] He introduced the prevailing urban planning discourse—one that marginalized cycling—into the Cyclists' Federation.

The planning department's proposal was, however, not completely unchallenged in the city council. The Social Democrat Harry Ljung criticized the one-sided focus on car traffic and drivers. Removing the bicycle lanes implied, according to Ljung, creating an "autostrada in the midst of the city," and he ridiculed the Cyclists' Federation for supporting the proposal. Ljung, citing the daily cycling commuters of the southern suburbs, argued that while many used the bicycle for exercise, for most people it was an economic matter:

> Their economy will not allow them to use trams and buses, and instead they employ their muscle power when they travel to and from work.… I believe that the city must also build for this category of citizens, which, for the most part, consists of people who build our houses, produce our machines, work in offices and warehouses; that is, the really useful and productive citizens of our city.[21]

Most importantly, Ljung and others reacted to the way the scrapping of the bicycle lanes was justified by statements about the expected decline of bicycle traffic. He argued that "cycling frequency depends entirely on whether we create facilities for cyclists. If we do, we encourage people to ride bicycles."[22]

Turning to Essingeleden, a large traffic route with bicycle lanes over the Essinge islands was included in the 1952 Master Plan for Stockholm.[23] In 1957, the urban planning department proposed a bridge with four lanes for motor traffic in a first phase, which could later be expanded to six lanes plus dedicated lanes for pedestrians and cyclists. According to a new proposal approved by the city council in 1959, however, the route should be designed with *at least* six lanes. The winning proposal in 1961 included eight lanes from the start.[24]

While the number of car lanes increased during the planning process, the 1959 decision included scrapping the pedestrian and bicycle lanes on Essingeleden. The traffic department and the Council of Mayors found a need for these lanes. The National Road Administration, the authority responsible for the road, however, opposed the lanes, claiming that they would take too much space and the road would not have a capacity that matched the expected volume of (car) traffic. Running the risk of losing state funding for 95 percent of the construction costs of Essingeleden, the city council found little choice but to follow the recommendation of the National Road Administration.[25]

The case was not entirely closed, however; in 1963, the bicycle lanes were up for discussion once more. Through their contacts with the traffic commissioner, Social Democrat Helge Berglund, the Cyclists' Federation had high hopes that Essingeleden would be equipped with bicycle lanes after all. In March 1964, however, alarmed by press reports stating otherwise, the organization turned to Berglund and the traffic board to express their concerns. They urged the city to consider all possibilities to equip Essingeleden with bicycle lanes, at least on Gröndalsbron, the section between Gröndal on the southern mainland and Stora Essingen, an island in Lake Mälaren. Since the route served major industrial areas south and west of the city center, scrapping the bicycle lanes would mean that working-class cyclists had to make a major detour on their way to and from their jobs in these areas. Moreover, the costs involved in adding bicycle lanes would be negligible in relation to the cost of constructing the route in its entirety.[26]

The traffic board, advised by the traffic department, supported the proposal to scrap the bicycle lanes, however. Essingeleden was part of the 1960 highway plan, the department wrote, whose purpose was to produce "a widespread network of high-capacity highways." Berglund's communications with the traffic director, moreover, reveals that his hands were tied by an agreement with the National Road Administration in which he had traded the bicycle lanes in return for entrance and exit ramps to Gröndal, an area adjacent to the route.[27]

The traffic department developed arguments about why cyclists should not be allowed access to the Essingeleden. It would be "very unsatisfactory ... from a safety point of view" to allow bicyclists on the shoulder of a highway with the high speeds expected there. Moreover, to have cyclists pass through the already very complex interchanges would pose "technically and economically unreasonable traffic solutions." It was theoretically possible to add bicycle lanes just to the Gröndalsbron section of the road. Such an arrangement would, however, involve multiple inventions: one car lane had to be abolished in each direction, the ramp for approaching car traffic would become more congested, and the ramps would still have mixed traffic. As this section of the highway was expected to become the busiest of them all, and its capacity was thus extremely important for the capacity of the entire route, it would be a clear disadvantage to reduce the number of lanes.[28]

Only one politician, Liberal councilor Charles Carlberg, objected to scrapping bicycle lanes on Essingeleden and other main roads that were part of the highway plan of 1960. Bicycle lanes along Essingebron would be welcomed, he argued, by people living in the southwestern suburbs of Stockholm who would otherwise see their conditions for commuting by bicycle deteriorating (relatively speaking).[29] The former chairman of the Stockholm chapter of the Cyclists' Federation, Carl Ehrnfelt, was harsher in his criticism: to construct Essingeleden with eight car lanes and none for cyclists was, he argued, "inhuman." With such an arrangement it was not surprising that car traffic was increasing rapidly while cycling declined—in fact, car traffic "was chasing [the weaker bicycle traffic] away."[30]

The opponents to scrapping the bicycle lanes—few as they were—nevertheless brought attention to the fact that some cyclists were highly dependent on bicycles and had few alternative forms of transport; they argued that single-sided planning and investment in automobility was not only a reaction to a transition from cycling to driving, but actually helped bring about this transition. The cases of Centralbron and Essingeleden suggest that—although the configuration of cycling and driving was already shifting, and many people "voluntarily" opted for the car—the reconfiguration was both accelerated and frozen by the decisions to scrap bicycle lanes on crucial commuter routes. In a scientific and seemingly neutral discourse, the planners thus quickly dismissed cyclists.

Reevaluating Cycling in Car Society

Both Centralbron and Essingeleden were constructed as exclusive motorways, in accordance with contemporary planning documents. The plans met little resistance. According to the prevailing interpretations of the bicycle—as a local, dangerous, and recreational mode of mobility, not at all suited for home-to-work journeys in the functionally dispersed city—(working-class) cyclists would have little interest in using these traffic routes anyway. Even if other interpretations existed among other social actors, these actors did not have the political leverage needed to seriously question the (noncyclist) planners in charge of designing the postwar city.

Since the late 1960s—in the wake of growing attention to the negative effects of automobility in terms of safety, noise, and pollution and opposition to car-oriented urban redevelopment—the negative representations of the bicycle have been complemented by positive ones, and they have increasingly spread among more "relevant" social actors. While earlier criticism of urban automobility pointed to negative *side*-effects while still viewing automobility in general as highly desirable, in the late 1960s a broader resistance toward car-oriented urban reconstruction and highways emerged in Stockholm, spearheaded by local residents' associations and urban environmental groups, most famously Alternativ Stad. In the years around 1970, these groups took to the streets, partly in the form of bicycle rallies, and eventually managed to pressure local politicians into reevaluating the rationalist urban redevelopment and traffic plans of the 1960s.[31]

Today, the representation of cycling has changed significantly since the postwar period, and it is no longer seen as the vehicle of the poor; at the same time, urban automobility has become an increasingly contested issue. But the reconfiguration of cycling and driving continues to be held back by those configurations that were embedded in the material structures of the city during the postwar period. The "modernization" of the bicycle can be partly explained by its inclusion in popular green discourses as an environmentally benign technology and its standing as a lifestyle product associated with health and fashion.[32]

While the "greening" of the bicycle has certainly reconceptualized it as a more modern form of mobility, studies carried out in Stockholm in the twenty-first century suggest that the choice to cycle is very seldom based on environmental considerations. Rather, the predominant motives for riding a bicycle are personal health and a view of the bicycle as the fastest and most flexible mode of transport in urban settings. Many people ought to consider cycling as a way to combine their daily commute with healthy exercise.[33] In addition to such practical concerns, environmental historian Wolfgang Sachs suggests there are also more abstract motives: in an (almost) fully motorized society, the bicycle has come to represent nonconformity and independence from traffic and congestion as well as the entire transport system.[34]

Indeed, British studies suggest that the reasons for choosing cycling were largely unchanged throughout the twentieth century. The bicycle is (according to cyclists) cheaper, faster, and simpler than other modes. In addition, environmental considerations have become a common motive in recent decades—although this factor is not as important for individuals as it is for policy makers.[35] What *has* changed is the social and political position of cyclists and, consequently, who understood the bicycle in this way.

Already in the 1970s, the members of the early urban environmental groups, who appropriated the bicycle as a symbol in their struggle against car-oriented urban redevelopment, were to a large degree young members of the educated middle class, protesting against the alleged atrocities of their parents' generation.[36] Studies show that today's cyclists in Stockholm are well-educated, above-average income earners. The vast majority has a driving license and they are relatively evenly distributed among the sexes. With a gross generalization, cyclists belong to a time- and health-conscious middle class.[37] Importantly—as the "renaissance of the bicycle" has been accompanied by a parallel "renaissance of the city," which is at the same time social, spatial, and economic—this is a group that has increasingly come into the spotlight in the postindustrial urban economy.[38] In other words, the new interpretations of cycling are currently held by well-positioned social groups.

Closing in on Stockholm, the urban environmental movement, the bicycle lobby, and later local politicians campaigned for more consideration to be given to cyclists, eventually leading to the first bicycle plan for Stockholm in 1978. Around the mid-1970s, the rapid postwar decline of cycling in Stockholm was reversed and followed by a modest increase. By the early 1980s, however, cycling was once more on the decline.[39]

The present upward trend began in the early 1990s, and has rapidly increased over the last ten years. Bicycles currently constitute 10 percent of all traffic passing in or out of the city center—indicating the high levels of bicycle commuting in particular. (Travel surveys capturing all journeys, on the other hand, show a relatively stable level of cycling around 5 percent.) The level of attention of policy makers matches the changes in bicycle use relatively well, the 1980s being a period of rather little interest in cycling compared to the 1970s. Interest increases again especially starting in 1998, when the local Stockholm Party managed to use their few mandates to urge the city government to substantially increase its commitment to improving conditions for cyclists.[40]

Meanwhile, providing cyclists with infrastructure in the form of a comprehensive network of dedicated lanes has become a ubiquitous policy solution for more and safer cycling. In dense urban areas, such as the city center of Stockholm, however, the process of establishing bicycle lanes is conflict laden. While some less-controversial bicycle lanes were established in the late 1970s, around 1980 the traffic board opted for what it called simpler and cheaper solutions—in particular painting "bicycle lanes" on wider pavements—supposedly to speed up the development of a coherent network of bicycle paths.[41] Explaining why prior investments in the bicycle network had been scaled down, Conservative traffic commissioner Sture Palmgren in 1985 referred to economic hardships, but acknowledged that this was not the only reason:

> Stockholm is to a large degree a completed city. Street space is confined by houses, water and bridges. The competition for this scarce space is stiff. SL [the public transport authority] wants bus lanes, commercial traffic wants possibilities for moorage, the residents need parking places, and traffic must keep moving to prevent traffic jams and cars from polluting the air. Since Stockholm is a big city and moreover the capital of a country with a large amount of people in motion every day; these needs cannot be neglected. To

lay claims on street space for bicycle lanes is therefore in many places technically impossible, in other places technically complex and therefore expensive. Given these circumstances I consider the development rate of bicycle lanes satisfactory.[42]

When conflicts of interest in urban traffic were brought to a head, bicycle-related measures had to yield. Since the 1990s, and especially since 1998, the bicycle network has been considerably extended, but often—in particular when bicycle lanes in major streets have been proposed—against strong opposition, not least from commercial traffic and the Stockholm Chamber of Commerce.[43]

As for the outer parts of the city and the conditions for commuting to the city center by bicycle, a new bicycle plan in 2005 drew attention to basically the same shortages that were identified in the first bicycle plan for Stockholm in 1978: the absence of a coherent and navigable bicycle network; that the existing cycle infrastructure was full of high curbs, barriers, and other physical obstacles; that cyclists on certain stretches shared space with pedestrians, and on other stretches with cars; and a low safety standard at the intersections with streets.[44] Decades had passed between the plans, but not much seemed to have happened. The future will tell whether the new bicycle plan adopted by the city council in spring 2013, with its pronounced focus on commuter routes for cyclists, will live up to its expectations.[45]

In spite of new interpretations of the bicycle (and automobility) since the late 1960s, the reemergence of cycling as a mass mode of transportation and a normal practice has been (until perhaps very recently) slow and discontinuous. Reconfiguration faces resistance from decades of car-centered planning and urban development, as well as the perceptions, habits, and routines that come along with it.

Reconsidering Past Decisions

The process of giving cyclists access to Centralbron and Essingeleden, or equivalents to them, has been slow and uncertain. Centralbron was in fact questioned from the start, as was the simultaneously (in 1947) adopted new railway bridge and the bridge for the subway over Riddarfjärden (the inlet to Lake Mälaren), not least from an aesthetic point of view and due to their impact on the historical buildings on the small island Riddarholmen.[46] The bicycle lanes were scrapped from Centralbron in 1951. When the southern part of Centralbron was opened to traffic in 1959, the planning process had brought about yet another change: in 1956, the city council had approved a wider bridge with room for six rather than the originally planned four car lanes.[47]

Traffic commissioner Helge Berglund, aware that bridges were a politically sensitive topic, provided extensive arguments as to why a widening of Centralbron was necessary. The bridge was "indispensable" to support "the immense traffic between the city center and the ever more populous southern parts of the city, where new suburbs were springing up at a pace exceeding anything seen before." Traffic engineers' prognoses pointed to an insatiable demand for space. According to Berglund, widening the bridge was of the utmost importance from a traffic point of view, while the aesthetic effect was only negligibly greater than that of a more narrow bridge. Thus, aesthetic objections were dismissed by Berglund as sentimental and unconstructive.[48]

A few years after the opening of the southern part of Centralbron, Liberal councilor Charles Carlberg found it unfair that cyclists were forced to detour while motorists' travel distance was considerably shortened. He asked the city council to rescind the decision to ban cyclists from the traffic route on a trial basis.[49] Both the traffic and police departments asserted that they had considered allowing cyclists on Centralbron (in mixed traffic) when it was opened, but in the end chose not to. While the traffic department highlighted safety reasons, the head of cabinet in the police department also acknowledged the importance of making full use of the capacity for car traffic: "The increasingly expanding metropolitan traffic demands for its existence broad and high-speed thoroughfares.... A prerequisite for the full use of the capacity of these traffic routes is, however, that they are not—other than through special lanes—open to vehicles that slow down faster traffic due to their limited speed capacity and general construction."[50]

Cyclists were thus never allowed on Centralbron.[51] They did, however, continue to ask for bicycle lanes—for example in the mid-1970s—while the traffic department continued to argue that eliminating a car lane for the benefit of cyclists was at odds with "overall traffic planning principles," which considered Centralbron the central north–south axis of the "inner city car traffic network." In addition, the entrance and exit ramps in the southern connection at Hornsgatan could not be satisfactorily designed to accommodate cyclists from a traffic safety point of view. Instead, cyclists had access to an overpass built just parallel to the railway bridge over Riddarfjärden in 1962.[52] That sidewalk, however, was clearly meant primarily for pedestrians; at its southern endpoint the bridge connected to Mariaberget, whose steep slopes and narrow cobblestone alleys makes it unattractive for cyclists—if they did not opt for taking the stairs or the elevator to the main street network. The midpoint connections to the bridge in Gamla stan, moreover, consisted of stairs only. When the Green Party proposed, in 2000, the construction of entrance ramps in Gamla stan, the traffic department found the proposal inappropriate as it would stimulate cycling over the bridge, increasing the risk of accidents and pedestrians' feeling of insecurity.[53]

Turning to Essingeleden, when Gröndalsbron (the section between mainland Gröndal and the island Stora Essingen) opened for traffic in 1966, the engineers designed it with a 1.3-meter-wide sidewalk. Carl Ehrnfelt, former chairman of the Stockholm chapter of the Cyclists' Federation, urged that cyclists should be allowed to use the sidewalk on the eastern side of the bridge. The traffic department was reluctant, however, since the narrow sidewalk was hardly wide enough for cyclists to navigate.[54]

Nevertheless, cyclists frequently used Gröndalsbron. During the following decades, repeated requests were made to widen the walkway and give cyclists formal access to it. In 1989, two Communist councilors wanted to narrow the car lanes on the bridge to give room for a wider combined lane for pedestrians and cyclists.[55] In 1996, a representative of the Stockholm Party wanted to create a bicycle lane by making use of the roadside space between the roadway and the walkway or by adding an additional ramp, noting: "Essingeleden was built at a time when one believed that in the future mankind would move about by car from cradle to grave."[56] And in 2000, a Leftist councilor suggested—based on a counter report written by a party colleague on the county level—"safe and high-quality walking and cycling paths" by transferring space from the car lanes of Essingeleden, which was "completely designed according to the needs of motorists."[57]

In response to the 1989 request, the traffic department considered the costs involved to change the complicated bicycle ramps for getting on and off the bridge and to maintain it during the winter prohibitive. Instead, the department and the majority of the traffic board put its hopes on having the bicycle lanes included in the light rail project between Gullmarsplan south of the city and Alvik on the mainland to the west that was being negotiated by the city and the county as part of a major traffic infrastructure investment plan for the Stockholm region, and which would parallel Gröndalsbron on bridges between Gröndal and Stora Essingen and further on to Alvik.[58]

In 1996, the traffic board, advised by the traffic department, first approved a widening of the sidewalk at the expense of the shoulder—only to find out that the National Road Administration had plans to accommodate the expected traffic increase on Essingeleden by making use of the same shoulder.[59] In 1999, fueled by the closure of a deal regarding a pedestrian and bicycle path along the light rail bridge between Alvik and Stora Essingen, the city initiated negotiations with the National Road Administration in order to widen the sidewalk on Gröndalsbron by making use of the shoulder. The widening was realized the year after. In return, the National Road Administration won the right to make certain changes regarding the entrances and exits in order to enhance the (car) capacity of Essingeleden.[60]

Almost thirty-five years after its construction, the narrow walkway on Gröndalsbron was finally widened in 2000 in order to give cyclists access to

Figure 5.2. In the homemade report "Sagan om den lilla, lilla gångbanan på den stora, stora bron" ("Fairy tale about the tiny, tiny walkway on the big, big bridge"), Leftist county councilor Jan Strömdahl designated Essingeleden one of the most distinct symbols of the car society of the 1960s. Even though cyclists were officially not allowed on the walkway, they made frequent use of it since it shortened many journeys considerably. Two cyclists could however barely meet on the narrow walkway without dismounting their bicycles. Photo by Jan Strömdahl.

one of very few north–south connections in Stockholm. Cyclists still lack access to Centralbron or an equivalent in central Stockholm, more than fifty years after its opening. Current proposals, however, do include a complementary bicycle connection over Riddarfjärden.

Conclusion

Since there are few connections between the northern and southern parts of Stockholm, access to them is crucial for commuters regardless of their means of transportation. The connections built during the 1950s and 1960s, while initially planned with bicycle lanes, were constructed with no such lanes, thus effectively excluding bicyclists from some of the quickest routes in the city. Repeated proposals to give bicyclists access to (parts of) the bridges has been, until very recently, in vain (allegedly because this was too costly, too difficult, too dangerous, too uncomfortable). These structures are "monuments of unsustainability": remnants (and symbols) of past regimes that continue to hold back more sustainable mobility patterns.

In this chapter, I have used the examples of Centralbron and Essingeleden to illustrate how infrastructures that were planned and built based on the postwar configuration of cycling and driving continue to resist reconfiguration according to new appraisals. The two structures have provided focal points of what is a more general feature in urban infrastructures. Long after the emergence of new interpretations of the bicycle, past representations, manifested in the urban infrastructure, continue to frame the conditions faced by cyclists navigating the city—as well as our perceptions about what is possible and appropriate by bicycle.

What is more, the "renaissance" of the bicycle is ambiguous; it not only confronts the remnants of a past car-oriented regime, but is also paralleled by a continuous reproduction of that regime. In Stockholm, reurbanization and increased cycling in the inner city and just beyond is accompanied by continued dispersal, longer journeys, and decreasing cycling in the region.[61] Moreover, the overarching infrastructure plans, which cater to automobility, have proven hard to kill. Slimmed down in the 1970s, an agreement about a great road (and public transport) scheme was finalized in 1991. While it was successfully obstructed in the short term by a quickly mobilized urban environmental movement, the separate links have recently been realized or are likely to be in the near future.[62] Finally, next to an investment plan of one billion Swedish kronor for cycling over a ten-year period, more funds than ever are spent on infrastructure projects for automobility.[63]

The postwar interpretations of cycling and driving continue to determine larger traffic investments. There are many complementary reasons for this, ranging from straightforward resistance to change from powerful groups and the accumulated investments in infrastructures, to institutional arrangements and established traditions and approaches to solve urban transport problems.[64] Not least important for the obduracy of the car city in the face of new, sustainable ideals are the powerful beliefs in freedom of movement and the ability to bridge distances that have shaped postwar cities; beliefs that are reproduced by urban sprawl and the functionally separated city—the spatial forms of "modernity" in the welfare state. In spite of increasing congestion and traffic jams, for many people the car is still considered the only way to satisfy the travel needs that it helped to shape.

Martin Emanuel holds a PhD in the history of technology (2013) from the Division of the History of Science and Technology at the Royal Institute of Technology (KTH) in Stockholm. He is currently a postdoctoral researcher at the STS Center and the Department of Economic History at Uppsala University and a visiting fellow of the Technology, Innovation, and Society Group at the Department of Industrial Engineering and Innovation Sciences, Eindhoven University of Technology.

Notes

The research for this chapter was carried out within the interdisciplinary research program CyCity and funded by VINNOVA, Sweden's innovation agency.

1. Samuel Strandberg, "Kursändring hur fort?" in *Gang-, cykel- og knallerttrafikken som integreret del af transportudviklingen: Foredrag m.v. ved NKTF:s konference den 16–18 april 1980 i København* (Copenhagen, 1980), 3.
2. Ibid., 3–4.
3. Depicting bridges as monuments of modernity is common. For example, in introducing the theme of his book, Richard Dennis starts with the construction and representation of the Brooklyn Bridge (New York), Tower Bridge (London), and Bloor Street Viaduct (Toronto), noting, for example, that the Brooklyn Bridge represented "technological modernity ... segregation of users and ... efficiency in circulation" and that it quickly assumed "the status as an icon." Richard Dennis, *Cities in Modernity: Representations and Productions of Metropolitan Space, 1840–1930* (Cambridge, 2008), 4–20.
4. Sustainability is here understood in terms of the UN Brundtland Commission's definition of sustainable development as "development that meets the needs of the present without compromising the ability of future generations to meet their own needs." Regarding sustainable mobility specifically, an attempt to define the term has been made by David Banister. Among the many elements for achieving sustainable mobility, a modal shift toward more walking and cycling is particularly important. David Banister, "The Sustainable Mobility Paradigm," *Transport Policy* 15, no. 2 (2008): 72–80.
5. Elizabeth Shove, "The Shadowy Side of Innovation: Unmaking and Sustainability," *Technology Analysis and Strategic Management* 24, no. 4 (2012): 363–375. See also Adri

Albert de la Bruhèze and Frank Veraart, *Fietsverkeer in praktijk en beleid in de twintigste eeuw* (Eindhoven, 1999).

6. For an overview, see Colin McFarlane and Jonathan Rutherford, "Political Infrastructures: Governing and Experiencing the Fabric of the City," *International Journal of Urban and Regional Research* 32, no. 2 (2008): 363–374.
7. Steven Lukes, *Maktens ansikten* (Stockholm, 2008), 115–157.
8. James Mahoney, "Path Dependence in Historical Sociology," *Theory and Society* 29, no. 4 (2000): 507–548, 507.
9. For an illuminating discussion on obduracy in the city, see Anique Hommels, *Unbuilding Cities: Obduracy in Urban Sociotechnical Change* (Cambridge, MA, 2008).
10. Arne Dufwa, *Stockholms tekniska historia: Trafik, broar, tunnelbanor, gator* (Stockholm, 1985), 75–78.
11. Ibid. Cycling boomed in most West European countries during the interwar period. Albert de la Bruhèze and Veraart, *Fietsverkeer in praktijk en beleid in de twintigste eeuw.*
12. This and the following paragraphs are based on my dissertation: Martin Emanuel, *Trafikslag på undantag: Cykeltrafiken i Stockholm 1930–1980* (Stockholm, 2012). See also "Constructing the Cyclist: Ideology and Representations in Urban Traffic Planning in Stockholm, 1930–70," *Journal of Transport History* 33, no. 1 (2011): 67–91.
13. Shove, "The Shadowy Side of Innovation."
14. Ruth Oldenziel and Adri Albert de la Bruhèze, "Contested Spaces: Bicycle Lanes in Urban Europe, 1900–1995," *Transfers* 1, no. 2 (2011): 29–49; Thomas Fläschner, "Stahlroß auf dem Aussterbe-Etat: Zur Geschichte des Fahrrades und seiner Verdrängung in den 50er Jahren," *Eckstein. Journal für Geschichte* no. 9 (2000): 4–22.
15. Compare, for example, with the cities examined in Albert de la Bruhèze and Veraart, *Fietsverkeer in praktijk en beleid in de twintigste eeuw,* 14.
16. For a comparison of post–World War Two policies encouraging private automobility in the Scandinavian countries, see Knut Boge, *Votes Count but the Number of Seats Decides: A Comparative Historical Case Study of 20th Century Danish, Swedish and Norwegian Road Policy* (Oslo, 2006), 80.
17. See also Bernd Kreuzer, "1 Fahrrad=0,25 PKW-Einheiten: Das Fahrrad im Stadtverkehr zwischen verpassten Chancen und gewollter Marginalisierung, Pfadabhängigkeiten und Gestaltungsspielräumen," in *Erfahrung der Moderne. Festschrift für Roman Sandgruber zum 60. Geburtstag,* ed. Michael Pammer, Herta Neiß, and Michael John, 1 Fahrrad=0,25 PKW-Einheiten (Stuttgart, 2007).
18. The Stockholm city planning department argued for the scrapping in two memorandums by af Klercker and his colleagues. What follows is based on these memos if not otherwise indicated. Stockholms stads/kommunfullmäktiges handlingar [Documents of the Stockholm City/Municipal Council, hereafter SF] Utlåtande [Report] 428/1950, "PM angående cykeltrafiken på leden Södergatan–Tegelbacken," C. H. af Klercker, 6.10.1950; SF Utlåtande 13/1951, Stadsplanekontorets (C. H. af Klercker, Göran Sidenbladh and Sven Markelius) promemoria, 5.1.1951.
19. "Vår syn på nya Söderbron," *Cyklisten* no. 1–2 (1951): 14. Regarding the alternative cycling routes, see also SF Utlåtande 13/1951.
20. Carl-Henrik af Klercker, "Det framtida Stockholm i bilismens tecken," in *Samfundet Sankt Eriks årsbok 1956* (Stockholm, 1956).

21. SF Yttrande [Statement] 73/1951.
22. Ibid.
23. Carl-Henrik af Klercker, "Trafik- och parkeringsproblemet i Stockholm," *Svenska Kommunal-Tekniska Föreningens Handlingar* no. 6 (1952): 1–18, 13–17.
24. Rikard Skårfors, *Stockholms trafikledsutbyggnad: Förändrade förutsättningar för beslut och implementering 1960–1975* (Uppsala, 2001), 148–89.
25. Bo Ericson, Göran Johnson, and Gunilla Rudehill, "Ska vi cykla i Stockholm? Inventering och förslag till framtida cykelvägnät i Stockholms ytterstad, Solna och Sundbyberg" (MA thesis, Royal Institute of Technology, Stockholm, 1972), 29–30.
26. Trafikkontorets arkiv, Stockholms stad [Archive of the Traffic Department, City of Stockholm, hereafter TA], Gatukontorets arkiv [Archive of the Streets Department, hereafter GA], Dnr 1249/1964. Cykel- och Mopedfrämjandet (Thörnqvist) till borgarrådet Helge Berglund, 19.3.1964; Cykel- och Mopedfrämjandet (Thörnqvist) till Stockholm stads gatunämnd, 19.3.1964.
27. TA, GA, Dnr 1249/1964. Helge Berglund till gatudirektör Carl Ehrman, 24.10.1964; Gatukontoret (Ehrman/Dufwa) till Gatunämnden, 24.10.1964; Utdrag ur GN Protokoll 12.11.1964, § 36.
28. TA, GA, Dnr 1249/1964. Gatukontoret (Ehrman/Dufwa) till Gatunämnden, 24.10.1964.
29. SF Yttrande 447/1964.
30. TA, GA, Dnr 2573/1966. Ragnar Ehrnfelt till Stockholms stads Gatukontor, 17.7.1966; Gatukontoret (A Cronström/S Vikander) till Ragnar Ehrnfelt, 9.8.1966.
31. Emanuel, *Trafikslag på undantag*, chap. 8.
32. See, e.g., Manuel Stoffers's contribution in this volume. See also Dave Horton, "Environmentalism and the Bicycle," *Environmental Politics* 15, no. 1 (2006): 41–58; Zach Furness, *One Less Car: Bicycling and the Politics of Automobility* (Philadelphia, PA, 2010).
33. Ulla Ericsson, "Ökad cykelpendling, men hur? En undersökning om attityder till cykling bland boende i innerstadsnära bostadslägen" (Stockholm, 2000); "Att cykla i Stockholm: Så tycker stockholmarna" (Stockholm, 2006); Maria Börjesson and Jonas Eliasson, "The Benefits of Cycling: Viewing Cyclists as Travellers rather than Non-Motorists," in *Cycling and Sustainability*, ed. John Parkin (London, 2012), 260–265.
34. Wolfgang Sachs, *For Love of the Automobile: Looking Back into the History of Our Desires* (Berkeley, CA, 1992), 196–202.
35. Colin G. Pooley and Jean Turnbull, "Modal Choice and Modal Change: The Journey to Work in Britain since 1890," *Journal of Transport Geography* 8, no. 1 (2000): 11–24. William Steele suggests, for the Japanese case, a divide between policy makers and cyclists in this respect. See Steele's contribution in this volume.
36. Ulf Stahre, *Reclaim the streets: Om gatufester, vägmotstånd och rätten till staden* (Stockholm, 2010).
37. "Att cykla i Stockholm"; Maria Börjesson and Jonas Eliasson, "The Value of Time and External Benefits in Bicycle Appraisal," *Transportation Research Part A: Policy and Practice* 46, no. 4 (2012): 673–683.
38. Håkan Forsell, "Den kalla och varma staden: Stockholm som arena för migration och invandring i samtidshistorien: En introduktion," in *Den kalla och varma staden: Migration och stadsförändringar i Stockholm efter 1970*, ed. Håkan Forsell (Stockholm, 2008), 11–13.

39. Emanuel, *Trafikslag på undantag*, chap. 8 and page 346.
40. Ibid., 346–47; Erik Beckman and Svante Linusson, eds., *1000 meter cykelfält som skakade Stockholm* (Bromma, 2009).
41. TA, GA, Dnr 2475/1978. Utdrag ur GN Protokoll 19.4.1979, § 35; Gatukontorets (Brynell/Köhlmark) tjänsteutlåtande 17.4.1979, "Att anordna cykelbanor på breda gångbanor."
42. SF Protokoll 7.10.1985, attachment 9.
43. Emanuel, *Trafikslag på undantag*, 347.
44. "Cykelplan för Stockholms ytterstad" (Stockholm, 2005), 7.
45. "Cykelplan 2012" (Stockholm, 2012).
46. Gunnar Sandin, *Vägen till citybanan: Spårfrågan mellan Norr och Söder under 150 år* (Stockholm, 2012), 110f.
47. SF Utlåtande 268/1956; Protokoll 17.9.1956, 808; *Broarna över Söderström. Redogörelse över de nya järnvägs-, gatu- och tunnelbanebroarna mellan Södermalm och Gamla stan* (Stockholm, 1959), 19.
48. Helge Berglund, *Broar och stadsbild: Anförande av Borgarrådet Helge Berglund vid Stockholms Stadsfullmäktiges sammanträde den 17 september 1956* (Stockholm, 1956).
49. SF Motion 13/1961.
50. SF Utlåtande 6/1962.
51. SF Protokoll 22.1.1962, 6.
52. TA, GA, Dnr 2926/1974. Kopia av GN-protokoll 19.4.1974, § 44; Gatukontorets tjänsteutlåtande (Brynell/Köhlmark) 15.5.1979, "Cykelbana på Centralbron." See also Dnr 2075/1976.
53. TA, GA, Dnr 2000-670-3687. Per Bolund (MP) and Vivianne Gunnarsson (MP) till Gatu- och fastighetsnämnden, 31.10.2000, "Cykelramp på Centralbron och Tranebergsbrons cykelbana"; Utdrag ur GFN-protokoll 12.12.2000, § 12; Gatu- och fastighetskontoret tjänsteutlåtande 10.11.2000, "Cykelramp på Centralbron och Tranebergsbrons cykelbana."
54. TA, GA, Dnr 2573/1966. Ragnar Ehrnfelt till Stockholms stads Gatukontor, 17.7.1966; Gatukontoret (A Cronström/S Vikander) till Ragnar Ehrnfelt, 9.8.1966.
55. SF Motion 132/1989.
56. SF Motion 25/1996.
57. TA, GA, Dnr 2000-620-234:1. Ann-Marie Strömberg (V) till GFN 25.1.2000, "Sagan om den lilla, lilla gångbanan på den stora, stora bron."
58. SF Utlåtande 59/1990, Protokoll 14.5.1990, 13.
59. SF Utlåtande 219/1996; Protokoll 15.11.1996, 40; Yttrande 282; TA, GA, Dnr 2000-620-234:2. GFK Tjänsteutlåtande 27.1.2000, "Cykelbana över Gröndalsbron. Skrivelse från Ann-Marie Strömberg (V)."
60. TA, GA, Dnr 2000-620-234:2. GFK Tjänsteutlåtande 27.1.2000, "Cykelbana över Gröndalsbron. Skrivelse från Ann-Marie Strömberg (V)"; GFK Tjänsteutlåtande 13.5.2002, "Avtal med Vägverket om anslutning till cykelbana över Gröndalsbron"; GFN Protokoll 6/2002, 28.5.2002, § 12.
61. Lars Nilsson, "The Return to the City: Twentieth Century Urban Development in Sweden," in *Reclaiming the City: Innovation, Culture, Experience*, ed. Marjaana Niemi and Ville Vuolanto (Helsingfors, 2003), 58–59. Börjesson and Eliasson, "The Benefits of Cycling," 250–252.

62. Bo Malmsten, *Dennisöverenskommelsen: Förhandlingen, aktörerna, innehållet* (Stockholm, 1993); Ulf Stahre, *Den gröna staden: Stadsomvandling och stadsmiljörörelse i det nutida Stockholm* (Stockholm, 2004).
63. Per Lundin, "Att tänka om staden med historia: En introduktion till Bilstaden," in *Bilstaden: USA visade vägen*, ed. Uno Åhrén and Per Lundin (Stockholm, 1960, rev. ed. 2010).
64. Hommels, *Unbuilding Cities.*

 PART II

Intersections

 CHAPTER 6

Bicycling and Recycling in Japan
Divergent Trajectories

M. William Steele

Japan is one of the world's great bicycling nations, home to more than 86 million bicycle riders, and a society in which recycling has a long and established history. This chapter focuses on the history, present state, and future of recycling and bicycling in Japan.[1] My hope is that the case of Japan will inform discussions concerning the use of older technologies in helping to solve pressing problems that confront our common future. I conclude that while premodern patterns of thought and behavior, even in the form of an "invented tradition," can be used to achieve current and future environmental ends, the history of bicycling and the history of recycling have produced different results. Although the experience of recycling in premodern Japan has been remarkably useful in twenty-first-century sustainability campaigns, patterns of thought and behavior associated with bicycling have proven more ambiguous.[2]

Recycling in Japan

Beginning in the mid-1990s, Japan embarked on a national drive to reduce waste and increase recycling. The Ministry of Environment, upgraded to full ministry status in 2001, enacted a series of tough environmental laws designed to create a "sound material-cycle society" (*junkan-gata shakai*), based on the three Rs: Reduce, Reuse, and Recycle.[3] While the recent specter of global warming has much to do with these initiatives, recycling in Japan has a long history—a history that government officials and environmental activists have sought to utilize in a quest to make Japanese everyday society more environmentally friendly.

The Edo period (1600–1868), which immediately preceded Japan's takeoff into modern economic growth, saw the development of a sophisticated system of recycling in all areas, including human and animal waste. Confucian arguments against wastefulness, coupled with strict legal injunctions, made Japan into what Susan Hanley, an authority on daily life in the Edo period, described as "a resource efficient culture with a high level of physical well being."[4] Of course, as Ruth Oldenziel and Helmuth Trischler note in the introduction to this volume, the valorization of premodern sustainability often derives from so-called invented traditions. Nevertheless, Japan's particular circumstances produced unusually thorough systems of recycling and reuse during the centuries immediately preceding its modern takeoff.

For more than two hundred years, the Tokugawa shogun who ruled Japan during the Edo period enforced a policy of limited contact with the outside world. Like a giant Easter Island, Japan was in many respects a closed system in which people had to make do with limited resources. And yet, during the same period, characterized by social and political stability, Japan's population doubled from around 12 million people in 1600 to over 25 million by the middle of the eighteenth century, and continued to grow at a slower pace, reaching 32 million at the end of the era in 1868. Population increase, combined with limited access to outside resources, led the shogunal government to enforce frugal, resource-conserving policies. As Hanley concludes,

> The Tokugawa solutions to limited resources allowed the Japanese to reach a high level of civilization using a minimum of resources, and wherever possible, natural renewable materials.... The general principles of using less energy, fewer resources, and reusing and recycling let many more people participate in a higher standard of living than if the trend had been toward wasteful use of scarce resources.[5]

The treatment of human waste (urine and feces) is one example of Edo recycling. By the middle of the eighteenth century, Edo was the largest city in the world having a population of more than one million. It was also one of the world's healthiest and cleanest cities. Human waste was treated as a commodity, sold to farmers for fertilizer, with barely a drop wasted. Indeed, farmers and city dwellers often negotiated exchanges of human waste for a supply of fresh vegetables. There were also sophisticated systems of collection, refurbishing/reproduction, and resale of used clothing, paper, wood and metalwork, straw, and even ash. What could not be recycled became landfill in several land-reclamation projects. The "environmentally conscious society" that characterized the daily life of the Edo period was thus the result of government policy and strict enforcement combined with economic rationality. Over time it also became a habit of the heart.

The imperial government that replaced the Tokugawa shogunate in 1868 actively sought to "catch up with the West," especially through industrial and military buildup based on the aggressive use of domestic and imported resources. Programs of industrialization and militarization paid off: by 1900, Japan was an emerging industrial economy and a world military power. Military victory against China in 1895 gave Japan its first overseas colony, Taiwan. Another victory in 1905, this time against Russia, gave Japan the status of a great power. However, as with other industrial military states in the twentieth century, modernization brought with it problems of pollution and environmental destruction. As Tanaka Shozo, Japan's pioneer environmental activist, put it just before he died in 1913: "Electricity is invented and the world is thrown into darkness."[6]

For much of the twentieth century, both before and after the Second World War, Japanese military and industrial policies encouraged the indiscriminate use of resources. To be sure, at the individual level, moral suasion campaigns in the years of military expansion, and during the immediate postwar reconstruction, decried the evils of luxury and waste, demanding frugality, efficiency, thrift, savings, and of course recycling.[7] Nonetheless, when Japan achieved unprecedented levels of affluence in the 1960s, these injunctions were quickly forgotten. Successive years of double-digit increases in GNP were accomplished at the expense of water, air, and land purity. Despite its own green premodern history, modern Japan's disregard for the natural environment transformed it into a nation known for its pollution rather than for its natural beauty.[8]

Whereas Japanese consumers enthusiastically embraced the new possibilities of "throwaway" culture, by the 1970s, unprecedented levels of garbage had created unprecedented problems for local and national governments. Sophisticated recycling programs did not exist; at most, a truck drove slowly through residential neighborhoods to pick up used newspapers and cardboard in exchange for a roll of toilet paper. Garbage disposal primarily meant landfill. A series of islands in Tokyo Bay, such as Yumenoshima, literally "Dream Island," were built entirely out of garbage, but space began to run out. A "garbage war" in 1971 resulted in a shift to incineration—which saved space but only added to Japan's air pollution.

Beginning in the 1980s, cities and towns throughout Japan began regular collections of paper and glass in addition to existing categories of burnable and nonburnable garbage. Over time, new categories were added, including weekly recycling collection of glass bottles, aluminum and other metal cans, plastics, plastic (PET) containers, used clothing, used bedding, hazardous waste, and several categories of paper. By the late 1990s, spurred on by a lack of landfill space, worries about health risks caused by incineration, and the urgent need to reduce carbon dioxide emissions, the Japanese central government began to take recycling seriously. In 1997, it enacted the Law for the Promotion of Sorted Collection and Recycling of Containers and Packaging and in 2000 announced a plan to turn Japan into a "sound material-cycle society." As one government report put it:

> Mass production and mass consumption tend to lead to a mass-disposal society.... Conventional socioeconomic activities are also closely related to the exhaustion of natural resources, the destruction of nature, and the disruption of sound material cycles in the natural world. These activities and the climate change and ecosystem crises feed on each other in a vicious circle, resulting in increasingly serious global environmental problems. In light of the current situation, there is an urgent need for Japan as well as the rest of the world to establish a sound material-cycle (SMC) society based on reduced consumption of natural resources and lower environmental burdens. Such a society can be achieved by stepping up efforts toward sustainability in order to create a low-carbon society and a society in harmony with nature.[9]

Interestingly, this report took the form of advice from Hokusai, an artist active in the late Edo period, resurrected to help people in the early twenty-first century understand the fundamentals of a sound material-cycle society. Referring explicitly to the practices of the Edo era, the Japanese government has used history as the underpinning of its environmental policy. Hokusai even recommended spreading Japanese waste management technologies to other countries in East Asia: "Japan should access country-specific situations and needs and help Asian countries to improve their ability to use and process CRs [circulative resources] appropriately. That would enable them to put the 3Rs into action and properly dispose of waste."[10]

Government efforts have been reinforced by a popular interest in the history of the Edo period. In recent years, Ishikawa Eisuke, the author of several best-selling books on Edo period recycling and sustainability, has reminded his contemporaries that Japan was not always the throwaway culture it had become in the second half of the twentieth century.[11] Ishikawa argues that people in the Edo period recycled goods and materials because their supply was limited. Moreover, they had a different set of values: "Just three or four decades ago, Japanese culture valued minimizing consumption. This value was embodied in the concept of *mottai-nai,* meaning 'don't be wasteful.'" Ishikawa's argument is that, as Japan modernized, Japanese people rejected their own traditions in favor of policies that gave their country great power status but at the same time created environmental problems unknown in the days when they lived "a simpler life."[12]

Invented tradition or not, the use of history for environmental ends has had impressive results. According to a 2010 White Paper, Japan's resource productivity (an index that indicates how well industry and people effectively use materials) was 361,000 yen/ton in fiscal year 2007, an increase of 37 percent over fiscal 2000. The recycling rate (indicating the proportion of recycled and reused materials) was 13.5 percent in fiscal 2007, an increase of 3.5 percent from fiscal 2000. Unrecyclable waste, destined for landfill, was 2.7 million tons in fiscal 2007, a decrease of 53 percent from fiscal 2000.[13] In the same period, daily per capita household garbage decreased by 10.4 percent, and perhaps most impressive was the 77 percent decline in industrial waste.[14] The role of history has helped to revitalizing embedded environmental notions and practices. Interactions between old and new technologies and old and new ideologies and patterns of behavior have produced positive outcomes for the Japanese environment. Japanese people now take pride in their status as a recycling country.[15]

A Short History of Bicycles in Japan

The bicycle has followed a different historical and ideological trajectory. The first two-wheeler appeared on the streets of Tokyo in the revolutionary year of 1868.[16] The velocipede, or boneshaker, was an immediate curiosity. Other "wheels" followed, the result of Japan's new commitment to Westernization: horse-drawn carriages and carts, rickshaws, and locomotives. The rickshaw in particular served to bring about a revolution in mobility. It was a Japanese invention that quickly spread throughout Japan and to other parts of Asia. By 1872, three years after the new vehicle had been put into use, there were 40,000 and by 1875 over 100,000 rickshaw on the streets of the new capital. The number reached a peak in 1896 with 210,000 countrywide.[17]

The convenience of the rickshaw delayed interest in self-propulsion. The boneshaker, and from the 1870s, the ordinary or penny-farthing bicycle, attracted the attention of Japan's elite and served as a sort of status symbol. Nonetheless, the bicycle remained too uncomfortable, too unreliable, and too expensive to attract widespread use.[18] By the end of the 1870s, there were perhaps fewer than two thousand bicycles in Japan, some produced domestically, but most imported. The appearance of the safety bicycle in 1889, however, significantly altered the prospects of bicycling in Japan. Quickly, the bicycle had gained a reputation for speed and convenience. A bicycle manual published in 1896 noted marked improvements (*kairyō*) in the quality and dependability of bicycles, producing four major advantages to be derived from their use: speed (bicycles were faster than horses and rickshaw and on level roads were nearly as fast as a locomotive), health (bicycle riding provided excellent exercise, strengthening the body and clearing the mind), convenience, and economy (bicycles allowed people to save time and money).[19]

Moreover, in the early twentieth century, Japanese manufacturers such as Miyata took advantage of mass-production technologies, making for affordable bicycles and allowing Japan to experience a bicycle boom similar to, but by no means on the same scale of, that experienced in Europe and the United States. The number of bicycles registered in Japan was around 10,000 in 1890.

From that point on, bicycle ownership began to rise sharply. By 1901, there were 57,000 bicycles and by 1912 the number of two-wheelers had surpassed the 400,000 mark. By 1925, there were 4 million bicycles in Japan, doubling to 8 million in 1940.[20]

By the early twentieth century, the bicycle was clearly associated with the working classes, especially in urban areas. A 1901 letter to the *Chūō kōron* magazine called the bicycle "the best means of transportation for the common person" (*heimin-teki yūitsu no kōtsu kikan*), citing the economic advantages of its speed: "Time is money" (*taimu izu munee*).[21] The *Asahi* newspaper of 21 August 1902 described the streets of Tokyo as crowded with the daily traffic of over twenty thousand government officials, office workers, and delivery boys on bicycles, and declared that "the bicycle is no longer a luxury item."[22] Attracted both by utility and economy, ordinary people began to depend upon bicycle transport.[23] The number of bicycles overtook the number of rickshaws in 1908 and continued to expand. By the 1920s, bicycles were responsible for making the modern culture of speed, convenience, and mobility an everyday experience. In 1927, boasting the use of over 5 million bicycles, the Minister of Commerce and Industry declared 1 November to be Bicycle Commemoration Day, noting that Japan had become the "foremost bicycle country in the world" (*sekai dai-ichi jitensha koku*). The minister recognized the bicycle as the "speedy footwear of the working class" (*sankin kaikyū no sokudo no hayai geta*) and further noted the importance of the bicycle as an export item that generated over 200 million yen in revenue.

In the late 1930s, the bicycle was placed into wartime service, both on battlefields on the Asian mainland and on the home front. Bicycle ownership rose to nearly 8 million in the 1940s, but by the end of the war, numbers had significantly declined. In 1945, there were only 5.6 million bicycles in use, many in poor repair. As the postwar Japanese economy recovered, the number of bicycles increased rapidly. The 8 million mark was equaled in 1948 and rose in 1950 to 10 million. By 1960, the number was 20 million, growing to 30 million in 1970.[24]

In the early postwar years, bicycle usage, despite advances made by women, continued to be dominated by men. Threatened by increased use of scooters, motorcycles, and automobiles, bicycle manufactures began to actively target women. In 1956, Yamaguchi Bicycle put its Smart Lady model on sale. This sturdy but light one-speed unisex model, with cargo basket in front, was sold on a monthly installment plan. The pioneer of what came to be known as the *mama-chari* ("mother's chariot"), the Smart Lady was the easy-to-ride, inexpensive city bike that would transform Japan—men, women, and even children—into a nation of bicycle citizens.[25]

By 1968, the percentage of women in their twenties who declared themselves to be bicycle riders had risen to 90 percent.[26] In the ensuing decades, the commitment to Japan's "speedy feet" continued despite a dramatic rise in automobile ownership.[27] Currently, in the first decade of the twenty-first century, Japan ranks alongside other "bicycle nations" such as the Netherlands, Germany, and Denmark both in terms of per capita bicycle ownership and use. According to 2005 data, there were more than 86 million bicycles being used by a Japanese population of around 127 million. At 68 bicycles per 100 people, bicycle ownership in Japan ranks sixth in the world, following the Netherlands (109/100), Germany (85/100), Denmark (78/100), Norway (69/100), and Sweden (68/100); the United States ranked eleventh (44/100) and China seventeenth (31/100).[28] Data on Japanese bicycle usage is also remarkable, with bicycle trips comprising some 15 percent of all daily outings, ranking third in a comparison of twelve countries (the Netherlands ranked first at 30 percent; the United States was last at 1 percent).[29] Moreover, reflecting increasing energy costs and environmental concerns, there has been a significant shift from automobile to bicycle during the first decade of the twenty-first century, and especially after the 11 March 2011 earthquake and tsunami disaster.[30] Commuting rates are up in major urban areas, and sales of sport and specialty bicycles, high-tech electric-assisted bicycles or e-bikes, bicycles for mothers with two children, and bicycles for the elderly have risen sharply.[31]

Bicycles Abandoned

From the outset, bicycle advocates stressed the contribution of bicycles to the positive aspects of the modern experience. In addition to speed, convenience, and independence, bicycles were linked with progress, equality, individualism, democracy, and social and political reform as well as with innovations in technology, industry, and advertising. Moreover, bicycle riding was recognized for its contribution to physical and mental health. But it was only in the 1970s, influenced by some of the bottom-up movements mentioned in the introduction of this volume, that bicycles in Japan were championed as a nonpolluting and sustainable form of transport. The term *bikoroji* (bikecology) first appeared in 1972 when twenty-one bicycle groups established the Society for the Advancement of Bikecology, hoping thereby to propagate the idea that bicycles can contribute toward the creation of a "safe and comfortable environment." The current website of the society praises the bicycle as a nonpolluting and environmentally friendly vehicle.[32] Later, in 1998, some Japanese parliamentarians established the Japanese Legislators' Association for Promoting Bicycle Use. This group made a remarkable appeal to the 2002 Johannesburg Summit on Sustainable Development, declaring: "The Bicycle will Save our Planet."[33] Beginning around 2000, other eco-cycle NGOs, NPOs, and associations were founded linking the bicycle with environment, health, and economy.[34] In Japan as elsewhere, the bicycle, invented in the early nineteenth century, has emerged in the twenty-first century as the vehicle of environmentalism. Rising gasoline prices, moreover, have expanded bicycle use, making it not only environmentally friendly, but economically friendly as well.

Despite such accolades, it is questionable how deeply this new environmentalism has permeated the thought and behavior patterns of Japan's 86 million bicycle riders. A 2004 government-sponsored questionnaire of bicycle usage revealed that the overwhelming reasons in favor of bicycle riding were freedom of movement (66 percent), followed by speed (61 percent), economy (31 percent), health (30 percent), convenience (19 percent), enjoyment (12 percent), and finally, in last place, environmental friendliness (10 percent). A clear disconnect can by seen in the answers of municipal officials on reasons for encouraging citizen bicycle use: 84 percent praised the bicycle's contribution to the environment, 33 percent thought the bicycle would help ease traffic congestion, 24 percent cited health, and 1.5 percent saw the bicycle as cost effective.[35] Later surveys produced similar results. In 2010, for example, the city of Sakai administered a traffic questionnaire as part of an initiative to achieve status as a "model environmental city." In answering why people choose to ride a bicycle, the reasons given were economy (57 percent), health (51 percent), speed (30 percent), convenience (30 percent), environmental friendliness (23 percent), and high price of gasoline (5 percent).[36]

The mental framework informing the early use of bicycles (freedom, speed, convenience) has remained strong despite the active promotion of environmentalism by government and nongovernment agencies. Moreover, these early bicycle-riding patterns of behavior were not always socially acceptable. Immediately after its introduction to Japan, the speed of the bicycle was associated with nuisance (*meiwaku*), the promise of freedom (*jiyū jizai*) was met with demands for regulation (*torishimari kisoku*), and the push toward domesticity was countered by charges of ignorance of conventional morality (*dōtoku*). Bicycles may have been the "speedy feet" of the people, but they quickly competed for space with pedestrians, especially after 1970 when Japan's traffic laws were revised to permit bicycles to ride on pedestrian walkways.[37] Cyclists, particularly students, have been criticized for their self-willed and reckless behavior. Up to the present day, bicycle riders suffer from bad manners, bicycle traffic laws are routinely ignored, and bicycles are abandoned without regard to pedestrian traffic. Indeed, compared with more orderly bicycle nations in Europe, Japan is a sort of bicycle anarchy.

In particular, abandoned bicycles continue to plague cities large and small. Abandoned bicycles (in Japanese *hōchi jitensha*) refer to illegally parked machines (sometimes left for days in front of a store or on the side of the street close to a bus stop or train station), deliberately disposed machines (left in parks and other wooded areas or in areas normally out of sight), including

Figure 6.1. Abandoned bicycles in front of a "No Bicycle Parking" sign, Kichijōji, 1 May 2011. Photo by the author.

stolen bikes abandoned after misuse, and more narrowly, to illegally parked bicycles that have been impounded.[38] Nationwide, the number of abandoned bicycles is estimated to range between 5 million to 10 million bicycles each year, depending on the definition.

In the greater metropolis of Tokyo (population 13 million), the number of abandoned or illegally parked bicycles parked peaked in 1990 with an average of 243,000 each day.[39] As the capacity of parking facilities grew in the 1990s, the number of abandoned bicycles fell steadily. By 2001, the number stood at 200,000, but was half that figure by 2006. For 2010, figures show that some 686,000 people commuted by bicycle every day between their home and a train station; of these, 93 percent (638,00) parked properly in bicycle-parking facilities, but 7 percent, or some 48,000 bicycles, were literally abandoned in the vicinity of a station, much to the chagrin of shop keepers and pedestrians.

Local wards and cities within greater Tokyo recognize abandonment as a serious problem and have taken countermeasures, focusing on the creation of new and larger parking facilities. In 2010, for example, an additional 40,000 parking spaces were created, making for a total of 8,840,000 spaces available to Tokyo bicycle citizens. Special antiabandonment campaigns are carried out to enlighten bicycle riders to the importance of obeying traffic laws and proper parking practices. "No parking" signs are everywhere and sometimes special "manner police" are employed to patrol high-risk areas and direct bicycle riders to nearby parking facilities. At the same time, tougher measures have been adopted: abandoned bicycles are more frequently impounded and fines are imposed on their riders. In 2009, a total of 742,000 bicycles were impounded, down from a peak of 917,000 bicycles during 2006.[40]

Of these 742,000 bicycles, some 470,000 were claimed by their owners, leaving 314,000 bicycles for disposal. Municipal governments paid scrap agents to take 149,000 (47 percent) of these unclaimed bicycles, while 15,000 (5 percent) were passed over free of charge. The remaining 120,000 bicycles were sold to resource-collection agencies, of which 32,000 (10 percent) were recycled for sale. The number of recycled machines has remained fairly constant over the past five years, ranging between 32,000 and 35,000 bicycles.[41] The majority of these recycled bicycles were sold as used bikes within Japan, some (10,000) were sent overseas to developing countries, and some were reconditioned and used as rental bicycles as one means to counter the "bicycle pollution" caused by abandonment.

These costly policies (in 2009, local governments in Tokyo spent a combined total of 18.7 billion yen [around US 156 million dollars]) seem to have paid off. Overall, the number of abandoned bicycles is on the decline. As can be expected, however, there is much regional variation, both nationwide and within greater Tokyo itself. The western suburb of Mitaka (population 180,000), boasts the second largest number of bicycle commuters in Tokyo, with some

12,000 bicycles on their way to and from Mitaka station each day. A survey of traffic in front of the north side of Mitaka Station on 10 October 2008 revealed 6398 illegally parked and 6205 legally parked bicycles in the morning (between 11:00 and 12:00 AM), and 7074 illegally parked and 6712 legally parked bicycles in the afternoon (between 3:00 and 4:00 pm). The estimated average of abandoned bicycles—that is, bicycles tagged as abandoned—was around 600 each day. This number has not significantly declined over the past five years. Annually, the city impounds around 6000 to 7000 abandoned bicycles (see figure 6.2).[42] Of that number, about 60 percent are returned to their owners, leaving around 2000 bicycles unclaimed. Of these, only around 150 are recycled for reuse; the majority are sold for scrap. The number of bicycles registered in Mitaka in 2009 was 258,814, an increase from the 182,965 bicycles registered in 2005 (this includes bicycles no longer in use but left registered).

Figure 6.2. Mitaka City, Disposal of Impounded Bicycles, 2000–2009

Year	Impounded	Claimed	Unclaimed*	Return rate (percent)
2009	5628	3687	2082	66
2008	8157	5506	2752	67.9
2007	6849	4067	2528	60.3
2006	6334	3783	2864	60.5
2005	6416	3610	2971	57.9
2004	6131	3325	2725	55.4
2003	4887	3129	2009	57.6
2002	5441	2852	2430	54
2001	4713	2591	2311	56.6
2000	4631	2824		62.2

*Disposed bicycles primarily recycled as scrap in various categories according to conditions of disposal. Around 10 percent recycled for reuse.

Source: Mitaka City, "Summary of Administrative Affairs," 2001–2010.

Figure 6.3. Musashino City, Disposal of Impounded Bicycles, 2003–2007

Year	Impounded	Claimed	Unclaimed*	Return rate (percent)
2007	22147	14272	8527	64.4
2006	15373	8581	7238	55
2005	15808	7453	7453	47
2004	17075	7640	7640	44.7
2003	16955	9512	7398	56

*Disposed bicycles primarily recycled as scrap in various categories according to conditions of disposal. Around 10 percent recycled for reuse.

Source: Musashino City report on Urban Development, 2008.

Other cities have been more active in their campaigns to reduce abandonment and promote the recycling of unclaimed bicycles. For example, Musashino City (population 138,000) annually impounded twice as many abandoned bicycles as Mitaka, its larger neighbor to the south (see figure 6.3).[43] Toshima ward (population 715,544) in northwestern Tokyo is perhaps the most aggressive. In 2009, bicycle pickup crews went into action some 2778 times during the year, impounding some 41,180 abandoned bicycles. In addition to five major ad campaigns, the ward sponsored twenty-seven station-front campaigns warning of problems caused by illegally parked bicycles and of the consequences. Patrols went on duty 3001 times during the year and issued 108,134 warnings to would-be parking offenders. Finally, beginning in 2006, a bicycle tax on train lines operating within the ward was imposed, faulting their failure to provide adequate bicycle parking facilities.[44]

Bicycle Recycling and Its Limits

Since the 1990s, Japan has introduced recycling policies designed to create a sound material-cycle society in an attempt to drastically reduce waste. The initiative has been successful; everywhere in Japan people know their three Rs (reuse, recycle, and reduce). In this context, the recycling of abandoned bicycles is especially noteworthy. There are numerous bicycle-recycling shops, many of which are NGOs, and, more ominously, scrap-collection industries that actively seek to recycle (in one form or another) these abandoned vehicles. For example, Musashino City, Toshima Ward, and eleven other municipal governments throughout Japan regularly donate abandoned bicycles to MCCOBA (Municipal Coordinating Committee for Overseas Bicycle Assistance), an NGO that reconditions bicycles and sends them to developing countries. Since its founding in 1988, MCCOBA has sent more than sixty thousand bicycles to countries in Asia and Africa.[45]

Eco-Chari, set up in 2004, has made a successful business out of abandoned bicycles. It began by offering removal services to universities and other institutions suffering from "bicycle pollution."[46] Currently, some eighty-eight universities throughout Japan take advantage of its services.[47] Eco-Chari offers comprehensive bicycle recycling services: removal, repair, resale, and rental. It is a good example of a new green industry with a significant presence on the internet and on social networks such as Twitter. A final and much smaller example of bicycle recycling is the donation of some eighty abandoned Osaka bicycles to evacuation centers in Kamaishi City shortly after the 11 March 2011 tsunami. As one newspaper noted, roads were obstructed, cars and trucks destroyed, and gasoline in short supply, giving the eighty lucky bicycles a most welcome second life.[48]

The vast majority of abandoned bicycles, however, do not get such a second chance. According to a survey carried out the Japan Bicycle Promotion Institute in 2004, some 6,490,000 bicycles nationwide were slated for disposal. Of these, 920,000 (14 percent) derived from bicycle retailers who took in used or no longer functional bicycles, 1.1 million (17 percent) derived from unclaimed bicycles that had been impounded by local governments, and 4.45 million (69 percent) derived from waste-collection agencies, primarily operated by local governments. After processing these bicycles, some 7.5 tons of reusable metal was recovered from scrapping 5.53 million bicycles, a resource-recovery rate of 68 percent. A total of 660,000 bicycles were reconditioned for resale or for rental use, for a recycling rate of 10 percent. Finally, some 300,000 bicycles were compressed and used for landfill, making a loss of 22 percent, but still producing an impressive overall resource-recycling rate of 78 percent. More recent figures are unavailable, but two similar surveys were carried out in 1996 and 1998 with similar resource-recovery rates (69 percent and 78 percent, respectively).[49]

A recycling rate of 10 percent may seem low for a country that relies so heavily on two-wheeled transport. In fact, the low recycling rate and the large number of abandoned bicycles may be closely related. In the first decade of the twenty-first century, caused by growing environmental concerns, economic malaise and negative population growth at home, and the impressive takeoff of the Chinese economy, the makeup of the Japanese bicycle fleet has undergone an interesting transformation. The market for children's bicycles, mountain bikes, and minibikes had declined in favor of increased interest in sport and specialty bicycles and high-tech electric-assisted bicycles (e-bikes). In 2002, domestic production of e-bikes was only 7 percent of total production; by 2008, it had expanded to 25 percent of all bicycles produced in Japan. Nonetheless, the greatest demand continues to be light utility bicycles, the so-called *mama-chari*, most of which are inexpensive machines imported from China.[50] Thanks to these imports, it is now possible to purchase a new bicycle for around 10,000 yen (US 85 dollars). Cheap bicycles are disposable bicycles. Parking offenders find it easier to buy a new bicycle than to spend time and money on the paperwork and inconvenience of collecting their impounded vehicles. Several people compare these inexpensive bicycles to inexpensive umbrellas that can be thrown away after use.

It is precisely the poor quality of these disposable bicycles that makes their life expectancy short (less than five years) and their recyclability dubious. As Kobayashi Shigeki of the Japan Bicycle Usage Promotion Study Group has put it, "most abandoned bicycles are not worth recycling; they are of such poor quality that it is better simply to scrap them."[51] It may seem a great waste (*mottai-nai*), but he concludes that the poor design and low safety standards of the machines, coupled with long periods of outdoor storage, make them unsuitable for recycling.

Conclusion

The bicycle has justly been praised for its role in helping Japan reduce its carbon footprint. Recycling and bicycling are excellent examples of established technologies that are helping humans lighten the heavy burden they have placed on the environment over the past one hundred years. Consequently, when writing this essay, I had hoped to uncover parallel or interrelated paths. However, recycling and bicycling, rhymes aside, have constructed divergent histories, reflect different values, and intersect with environmentalism differently. History does not always repeat itself (even though it may rhyme).[52]

Recycling in the Edo period was enforced by a strong sense of obligation and an ideology that celebrated frugality and the simple life. These early values, strengthened by more modern moralizing campaigns and advice from Hokusai and others reminding people of past practices, have been revitalized to support contemporary recycling programs. However, the strict observance of law and internalized morality that advanced the recycling of household garbage has not had a similar influence on the observance of bicycle laws and manners. Although celebrated as a nonpolluting and sustainable mode of transportation, bicycles in Japan retain an unruly and subversive character. Introduced from the West in the 1870s, the bicycle came with a distinctive set of values: freedom, self-empowerment, speed, convenience, and liberation from the bonds of society. Those values have not been lost, as bicycles continue today to offer subtle resistance to the forces of social conformity. Here I suspect the history of bicycle riding in Japan is not unique.

There are those who dearly love their bicycles, but many people abandon them easily; they are, after all, unregulated, unregistered, underpoliced, and cheap. Moreover, bicycle design and manufacture do not lend themselves to easy disposal or recycling. Local municipalities are making progress in solving the problem, using a combination of convenience (additional parking facilities) and strong-arm techniques (policing and fines), but the will of cyclists to maximize freedom, speed, and convenience continues to supersede environmental concerns. Moreover, while tough recycling laws govern the "end of life" of automobiles and home appliances, there is no specific legislation for the disposal and recycling of bicycles.[53] Despite creative attempts by NGOs and businesses to recycle abandoned bicycles, ironically the "pollution-beating" bicycle is itself a source of pollution in Japan today.

M. William Steele is professor of modern Japanese history at International Christian University (ICU), Tokyo, Japan. He specializes in the social and cultural history of Japan in the late nineteenth century. He received his PhD from Harvard University in 1976. He is the author of *Alternative Narratives in*

Modern Japanese History (2003). His recent publications focus on mobility and environmental issues, including the history of rickshaws, bicycles, and automobiles in modernizing Japan.

Notes

1. This chapter is a revised version of a paper presented in "Re/Cycling Histories: Users and the Paths to Sustainability in Everyday Life," a workshop sponsored by the Rachel Carson Center, Munich, 27–29 May 2011. I wish to thank the conveners of the workshop for their encouragement to revise my talk for publication. Portions of the section on bicycle abandonment in Japan have been published in an International Christian University publication, "Bi-cycling and Re-cycling in Japan: History Does Not Always Repeat Itself," *Japan Studies: Frontiers* (ICU Japan Studies Program, 2013), 19–30.
2. As Ruth Oldenziel and Helmut Trischler point out in the introduction to this volume, despite the linguistic likeness, in Japan as elsewhere there has been little attempt to combine the history of bicycling and recycling. References below are given separately to literature on bicycle use and recycling in Japan.
3. Information on Japan's recycling initiatives may be found in several Environmental White Papers compiled by the Ministry of Environment, http://www.env.go.jp/en/wpaper/ (accessed 7 August 2011). Of particular interest is the 2010 review of the first ten years of Sound Material-Recycling Plan: "2010: Establishing a sound material-cycle society, Milestone toward a sound material-cycle society through changes in business and life styles," http://www.env.go.jp/en/recycle/smcs/a-rep/2010gs_full.pdf (accessed 7 August 2011).
4. Susan Hanley, *Everyday Things in Premodern Japan: The Hidden Legacy of Material Culture* (Berkeley, CA, 1999); see chap. 5, "Urban Sanitation and Physical Well Being," 104–128.
5. Ibid., 75–76.
6. Kenneth Strong, *Ox against the Storm: A Biography of Tanaka Shozo, Japan's Conservationist Pioneer* (London, 1995).
7. Sheldon Garon, *Molding Japanese Minds: The State in Everyday Life* (Princeton, NJ, 1986), 11.
8. On pollution in Japan, past and present, see Peter Kirby, *Troubled Natures: Waste, Environment, Japan* (Honolulu, HI, 2010). See also Jun Ui, ed., *Industrial Pollution in Japan* (New York, 1991); and the popular book by Alex Kerr, *Dogs and Demons: Tales from the Dark Side of Japan* (New York, 2002). Beginning in the 1970s, urged on by citizen protest movements, the Japanese government enacted strict environmental-protection legislation. The quality of Japan's land, air, and water resources has improved significantly over the past thirty years. See Hidefumi Imura and Miranda A. Schreurs, eds., *Environmental Policy in Japan* (Cheltenham, 2005).
9. Quoted from an informative White Paper issued by the Ministry of Environment that traces the developing of recycling from the Edo period up to the establishment of the Sound Material-Cycle Society legislation in 2000: *A Sound Material-Cycle Society through the Eyes of Hokusai,* available online: http://www.env.go.jp/recycle/3r/approach/hokusai_en.pdf (accessed 12 September 2014).

10. *A Sound Material-Cycle Society through the Eyes of Hokusai*, http://www.env.go.jp/recycle/3r/approach/hokusai_en.pdf (accessed 12 September 2014).
11. Ishikawa Eisuke, *O-Edo ekoroji jidai no jijō* (Tokyo, 2000), translated as *Japan in the Edo Period: An Ecologically Conscious Society*, by the NGO, Japan for Sustainable Society (JFS), http://www.japanfs.org/en/pages/009397.html (accessed 7 August 2011).
12. Ibid., chap. 2, "Darker Side of Convenience," http://www.japanfs.org/en/pages/022425.html (accessed 7 August 2011).
13. Technological advances in incineration aided in this result; it should be noted, however, that municipal waste incineration remains a major source of dioxin pollution.
14. Ministry of Environment White Paper, "2010: Establishing a sound material-cycle society."
15. Norimitsu Onishi, "How do Japanese Dump Trash? Let Us Count the Myriad Ways," *New York Times*, 12 May 2005, http://www.nytimes.com/2005/05/12/international/asia/12garbage.html (accessed 7 August 2011). See also, "Japan Streets Ahead in Global Plastic Recycling Race," *The Guardian*, 19 December 2011, http://www.theguardian.com/environment/2011/dec/29/japan-leads-field-plastic-recycling (accessed 12 September 2014).
16. For more details on the early history of the bicycle in Japan, see "The Speedy Feet of the Nation: Bicycles and Everyday Mobility in Modern Japan," *Journal of Transport History* 32, no. 2 (December 2011): 187–209.
17. On the history of the rickshaw in Japan and its spread to East and Southeast Asia, see M. William Steele, "Mobility on the Move: Rickshaws in Asia," *Transfers* 4, no. 3 (Winter 2014): 88–107.
18. Saitō Toshihiko, *Kurumatachi no shakaishi* (A Social History of Wheels) (Tokyo, 1997), 123.
19. Watanabe Shūjirō, *Jintensha-jutsu* (Bicycle Skills) (Tokyo, 1896), 10–11, available through the Japan National Diet Library Electronic Library, http://kindai.ndl.go.jp/BIBibDetail.php (accessed 7 August 2011).
20. M. William Steele, "The Making of a Bicycle Nation: Japan," *Transfers* 2 no. 2 (Summer 2012): 73.
21. *Chūō kōron*, June 1901, 78.
22. *Tōkyō Asahi shinbun*, 21 August 1902.
23. The appeal of utility and cost is the narrative strategy followed in major narrative history of the bicycle in Japan: *Jitensha no isseki – Nihon jitensha sangyōshi* (The Bicycle Century – History of Japan's Bicycle Industry), published in Tokyo in 1973 by the Japan Bicycle Promotion Institute (*Jitensha Sangyō Shinkō Kyōkai*). The role of military demand is also stressed. Two wars, the Sino-Japanese War (1894–1895) and the Russo-Japanese War (1904–1905), provided important stimuli for the Japanese bicycle industry. The bicycle proved useful for the military police, for reconnaissance, and for the delivery of messages on the battlefield. Accordingly, it was only after 1905 that the bicycle was pressed into the service of everyday life. For this argument, see Sano Yūji, *Jitensha no bunkashi* (A Cultural History of the Bicycle) (Tokyo, 1987), 166–172.
24. Steele, "The Making of a Bicycle Nation," 77.
25. For details on the history of the bicycle in postwar Japan, see ibid., 70–94.
26. Japan Bicycle Culture Center, *Tomo no kai tayori* no. 12 (October 2009), http://www.cyclo-info.bpaj.or.jp/japanese/index.html (accessed 7 August 2011).

27. Between 1966 and 2009, for example, automobile ownership jumped from 2.29 million to 57.68 million. For data on automobile and other motorized transport between 1966 and 2009, see the official site of the Japan Automobile Inspection and Registration Association, http://www.airia.or.jp/number/index.html (accessed 7 August 2011). During this same time, bicycle ownership rose from just under 30 million vehicles to the current 86 million.
28. According to data compiled on contemporary social, political, and economic trends (including Japan in comparative perspective) by the Honkawa Data Tribune, http://www2.ttcn.ne.jp/honkawa/6371.html (Per Capital Bicycle Ownership; accessed 7 August 2011).
29. Honkawa Data Tribune, http://www2.ttcn.ne.jp/honkawa/6370.html (International Comparison of Outings by Means of Transport, based on 1999 data; accessed 7 August 2011). Another remarkable set of statistics compares dependence on automobiles for transport with obesity; Japan (at the bottom of both scales) contrasts sharply with the United States (at the top of both scales). See Honkawa Data Tribune, http://www2.ttcn.ne.jp/honkawa/2240.html (International Comparison of Dependency on Automobiles with Obesity, based on 1999 data; accessed 7 August 2011).
30. *New York Times,* 18 April 2011, Special Report by Miki Tanikawa, "Out of Disaster, a Burst of Enthusiasm for Bicycling," http://www.nytimes.com/2011/04/18/business/global/18iht-rbog-bicycle-18.html?_r=2 (accessed 8 August 2011). See also "The Rise of Tokyo's Bike Commuters, or 'Tsukin-ists,'" 17 June 2010, in the Japan-Realtime section of the online *Wall Street Journal,* http://blogs.wsj.com/japanrealtime/2010/06/17/the-rise-of-tokyos-bike-commuters-or-tsukin-ists/ (accessed 8 August 2011).
31. See the 2009 report on a survey of bicycle-user needs issued by the Japan Bicycle Promotion Institute, *Riyōsha niizu ni motozuku jitensha no kaihatsu ni muketa chōsa kentō hōkoku* (Tokyo, 2009), 9–10. See also *Saikuru puresu Japan* no. 823 (February 2009): 20, for a report on the increase in demand for electric-assisted bicycles. In 2008, the number of electric-assisted bicycles exceeded 3 million. A recent survey examines the attitudes of men and women in their 50s and 60s toward bicycles: *Dankon no sedai ni taisuru jitensha no ishiki chōsa nado ni kansuru chōsa hōkokusho* (Tokyo, 2008).
32. The Society for the Advancement of Bicology, http://www.bikecology.bpaj.or.jp/guide/ (accessed 7 August 2011).
33. For their appeal, see http://www.cyclists.jp/legist/activities.html (accessed 7 August 2011).
34. See, for example, the Bicycle Usage Promotion Study Group, founded in 2000, http://www.cyclists.jp/ (accessed 7 August 2011).
35. Japanese Ministry of Transportation, "Toshi kōtsu ni okeru jitensha riyō no arata no kenkyu" (A Study of Bicycle Usage in Urban Traffic), 2005, http://www.mlit.go.jp/pri/houkoku/gaiyou/pdf/kkk58.pdf (assessed 7 August 2011). The results of the questionnaire on bicycle usage (by both riders and by municipal authorities) are on pages 12–14.
36. A study carried out by Sakai City, "Jitensha, teikōgaisha-nado ni kansuru shimin ishiki chōsa" (A Questionnaire on Citizen Attitudes toward Bicycles and Other Low-polluting Vehicles), 2010, http://www.sakaiupi.or.jp/09urban/vol22/22-13.pdf (accessed 7 August 2011). The answers regarding reasons for bicycle ridership are on page 82.
37. For details, see Motoda Yoshitaka and Usami Seiji, "Waga kuni ni okeru jidōsha-dō seibi ni kansuru rekishi-teki kōsatsu" (A History of the Maintenance of Bicycle Roads

in Japan), http://p-www.iwate-pu.ac.jp/~motoda/dobokukeikakuaki09motoda4.pdf (accessed 7 August 2011), 4.

38. On 31 January 2014, the *Asahi Shinbun* reported that some two hundred bicycles were pulled out of Inokashira Pond (Musashino City) when the pond was drained to clear it of invasive fish; http://www.asahi.com/articles/ASG1Y7FR2G1YUTIL046.html. See also *Japan Today*, "Abandoned bicycles dumped in Tokyo Pond," 27 January 2014, http://www.japantoday.com/category/national/view/abandoned-bicycles-dumped-in-tokyo-pond (assessed 12 September 2014).

39. Beginning in 1977, the Tokyo Metropolitan Government's Office for Youth Affairs and Public Safety has issued an annual municipal bicycle census, including the number of bicycles ridden to and from commuter stations (daily average), the number of available bicycle parking spaces, the number of bicycles legally parked (daily average), and the number of bicycles abandoned or illegally parked (daily average). The latest census (2010) is available on the Tokyo Metropolitan Government's homepage, http://www.seisyounen-chian.metro.tokyo.jp/koutuu/pdf/07_jitensyagenkyo2.pdf (accessed 7 August 2011). Data is derived from chart 1 (Trends in the number of abandoned bicycles, parked bicycles, availability of bicycle parking spaces, and number of bicycles used in commuting, 2010) on page 2 of the 2010 bicycle census.

40. Data derived from chart 3 (Trends in impounding, returning, and disposal of abandoned bicycles, 2009), on page 4 of the Tokyo Metropolitan Government's 2010 bicycle census (see note above). The chart covers the years between 1991 and 2009.

41. Data included on page 4 of the Tokyo Metropolitan Government's 2010 bicycle census (see note above).

42. Data on abandoned bicycles for the city of Mitaka may be found in *Jigyō no gaiyō* (Summary of Administrative Affairs), Mitaka: Toshi Seibibu, 2010, 87. Chart 1 (Mitaka City Disposal of Abandoned Bicycles) was compiled using back issues of this report beginning with the 2001 issue.

43. For details, see 2008 report of City Development compiled by Musashino City Office, http://www.city.musashino.lg.jp/cms/data/00/01/19/archive/11993-8.pdf (accessed 7 August 2011).

44. For details, see the Toshima Ward Office website, http://www.city.toshima.lg.jp/kotsu/jitensha/020294.html (accessed 7 August 2011).

45. For details on the activities of MCCOBA, see its website, http://www.joicfp.or.jp/eng/i_campaign/item/mccoba/index.shtml (accessed 7 August 2011). Toshima's relationship with MCCOBA is detailed on the Toshima Ward website, http://www.city.toshima.lg.jp/dbps_data/_material_/localhost/100doboku/030kotsuanzen/20kinenshi.pdf (accessed 7 August 2011).

46. See Eco-Chari's website for extensive information on its abandoned bicycle removal, recycling, and bicycle rental activities on university campuses, http://ecochari.com/newecochari/eco.html (accessed 7 August 2011).

47. Eco-Chari's website includes photos of before and after bicycle removal. See, for example, the case of Waseda University, http://ecochari.com/newecochari/disposalcase/dispo_waseda.html (accessed 7 August 2011).

48. *Osaka Nichi-nichi Shinbun*, 7 April 2011, http://www.nnn.co.jp/dainichi/news/110407/20110407031.html (accessed 7 August 2011).

49. Based on a 2005 survey carried out by the the Bicycle Promotion Institute, http://www.jbpi.or.jp/_data/atatch/2005/03/00000014_cyc_resources.pdf (accessed 7 August 2011).
50. In 2002, 72 percent of bicycles produced in Japan were light utility vehicles; by 2008, production had declined to 59 percent. This was accompanied by an overall decline in domestic production of bicycles; in 2002, there were a total of 3.076 million bicycles produced in Japan; this number fell to 1.095 million in 2008; in 2010, domestic production stood at 1.056 million. At the same time, the number of imported machines has increased dramatically. By 2010, more than 80 percent of bicycles purchased in Japan were Chinese imports, the majority being light, inexpensive utility bicycles. See the Data Archives of the Japan Bicycle Promotion Institute for a variety of statistics on production, imports, and exports of bicycles and bicycle parts, http://www.jbpi.or.jp/?sub_id=4&category_id=170&dir_no=TOP_ROOT:170 (accessed 7 August 2011).
51. Kobayashi Shigeki, "Kankō ni okeru jitensha no kanōsei ni tsuite" (On the Possibility that Bicycles can Promote Tourism," in *Kankō bunka* (Tourism and Culture), Special Issue (Bicycles and Regional Development) (March 2011): 3.
52. I thank Peter Cox for reminding me of Mark Twain's saying, "History does not repeat itself, but it does rhyme."
53. For the "End-of-Life Vehicle Recycling Law" and the "Home Appliance Recycling Law," see the Ministry of Economy, Trade and Industry website, http://www.meti.go.jp/policy/recycle/main/english/law/end.htmlhttp://www.meti.go.jp/policy/recycle/main/english/law/end.html (accessed 12 September 2014)

 PART III

Recycling Histories

 CHAPTER 7

Premodern Sustainability?
The Secondhand and Repair Trade in Urban Europe

Georg Stöger

It might not surprise us to learn that the premodern usage of objects differed significantly from today.¹ Many people took intensive care of their belongings—protecting, repairing, reusing, and finally recycling them.² Susan Strasser examined this "stewardship of objects" in her book *Waste and Want* for premodern and early industrial North America and insisted that this behavior should be "better understood not as a conscious virtue or as self-denial but as a way of life."³ Indeed, many household accounts, letters, pictorial sources, and contemporary literature indicate the omnipresence of such strategies of usage and related occupations.⁴ One could describe these forms of material usage and consumption as sustainable, as we can trace attempts to save resources and to enhance the life span of objects. Both aspects appear regularly in contemporary discussions on sustainability and sustainable development.⁵

In this chapter, I will investigate two fields linked to this premodern "recycling mentality," as Christian Pfister has called it⁶: secondhand trade and consumption as well as repairing. The work covers a period ranging from the seventeenth century until the first half of the nineteenth century and focuses predominantly on western and central European urban areas. First, I discuss impetuses for repairing and reusing and investigate how these forms of usage were linked to urban economy and consumption. Second, I examine which actors were involved and how contemporaries perceived them.

In the early twentieth century, German economist Karl Bücher, reconstructing patterns of labor in late medieval Frankfurt, highlighted what he called "developed repair crafts" and the importance of the "reworking of used goods."⁷ Since then, however, these occupations have not generated much interest among economic and social historians—the throwaway society of the twentieth century seems to have determined their points of view.⁸ Research has predominantly focused on aspects of production and (firsthand) consumption; "secondary markets"—i.e., those constituted by actors (traders, craftsmen, collectors, etc., as well as consumers) engaging in reselling, repairing, or recycling—have rarely been considered.[9] At present, there are only a few studies, mainly dealing with issues of premodern poverty or consumption, that consider the premodern secondhand trade, while repairing and recycling remain largely unstudied.[10] This disregard might also be a consequence of the difficulty of finding source material: due to their "oral" practice and their often informal nature, such businesses and transactions did not leave many written records. Furthermore, secondary markets often involved small-scale trade and craft—fields that were obviously considered irrelevant by late-nineteenth- and early-twentieth-century archivists, which resulted in a significant loss of archival documents. There is also a remarkable lack of newer empirical studies on the larger premodern crafts (such as tailors or shoemakers, who were important actors in the repair trade), on premodern (local) small-scale trade, on nonelite consumption, and on everyday informal economy, which currently limits our knowledge of secondary markets. Likewise, questions of premodern consumption and material usage (apart from the issues of energy and material flows) have not gained much attention among environmental historians so far.[11]

What caused large parts of premodern society to reuse, resell, repair, and recycle? Raw materials were usually expensive and often scarce, while labor was comparatively cheap.[12] In an economic sense, it was therefore reasonable to enhance the life span of objects, to reuse them, and to recycle materials. In addition, premodern people seem to have inherited a deeply rooted mentality of thrift that was not necessarily linked to their socioeconomic status.[13] Even in wealthier households, there are few signs of careless usage of commodities and materials.[14] But endemic poverty and scarcity of goods were obviously the primary triggers for such strategies, which can be seen as part of an "economy of makeshifts" of the urban laboring poor.[15] Secondary markets and product cycles helped to reduce household expenses, which offered few other possibilities to save money, since the cost of food could amount to half of the available earnings.[16] Material-saving strategies were also common within early modern craft, as shown in practices such as substituting material, economic use of materials, and especially reuse of waste materials.[17] Neglecting these secondary markets and secondary product cycles seems problematic: On the one hand, it can lead to an underestimation of the overall consumption and the output of an economy, when production is measured only by the input of primary material.[18] On the other hand, we might overlook important fields of economic activities and consumption. Into the industrial age, secondary markets were a means of livelihood for many people. For many craftsmen, especially in the textile branch, repair work formed a significant part of their daily business. Other townsfolk specialized in trading secondhand wares or in collecting and reprocessing scrap materials.

Repairing as a Household Strategy and an Occupation

Things that could be repaired were repaired—often for as long as possible. Repairing made good economic sense because of the relatively high material value of many commodities in contrast with the low labor costs. One example is clothing: Before industrial mass production, which developed and spread through continental Europe starting in the mid-nineteenth century, garments were relatively expensive. In the 1790s, a Salzburg periodical listed the price of a newly made jacket (*Rock*) for a rural female servant as 10 gulden (florins, hereafter fl.),[19] which was equivalent to more than thirty days of pay for a mason.[20] Only 12 kreuzer (hereafter kr.) were calculated for the tailor's wage. Other examples draw similar proportions between material and labor costs: In the Moravian town of Proßnitz (Prostějov, Czech Republic), an estimate for male prisoners' clothing (five jackets and five shirts) from the year 1786 noted 22 fl. 30 kr. as the cost for the fabric, and only 3 fl. as the tailor's wage.[21] In these two examples, the proportion of wage was 2 percent (Salzburg) and 10 percent (Proßnitz) of the total price. There are only a few indications of the costs of repairing garments. They were probably similar to the aforementioned wages. In 1770, the (wealthy) Salzburg merchant Franz Anton Spängler noted in his account book the expenses of 12 kr. for the mending of a jacket and 29 kr. for the mending and reworking of two pairs of trousers[22]; in 1789, a man from Salzburg paid 10 and 28 kr. to repair two pairs of shoes.[23] The decision to repair obviously depended on the value of the particular item (as well as on the availability of a replacement), but visual aspects seemed to matter as well, especially in the case of clothing.

Undoubtedly, repairing was a central household strategy among middle and lower urban strata; nonetheless, it was also widespread in wealthier households, as account books and other documents demonstrate. The merchant Spängler regularly documented expenses for repair work within his household in his account books, which span over five decades of the eighteenth century (1733–1785). Within this period, Spängler spent at least 200 fl. on the mending or reworking of clothing and shoes. Building repairs and the fixing of household goods are other recurring payments in the Spängler account books.[24] The correspondence of the Mozart family can serve as another example. Especially Leopold Mozart, Wolfgang Amadeus's father, frequently offered advice to members of his family on the necessity of mending clothes.[25] Moreover, repairing was central to semipublic urban institutions, such as work and poorhouses, orphanages, and the military, which tended to use inmates or persons linked with these institutions for repair work.[26]

Repairing was often combined with alterations, especially where clothing was concerned. Alterations were necessary to replace defective parts, but they also permitted people to adapt to changing fashions or liberate articles from their original context (for example, "newly" made clothing from used uniforms or servants' garments; see below), and they gave low- and middle-rank consumers the possibility of dressing more respectably or fashionably. Garments could be cut up entirely or dyed, or the fabrics could be turned over.[27] A Viennese satire from the 1730s mocks this practice: "one mends, one turns, one sews ... , one stitches everything together so that it follows fashion ... ; most things seem to be new, but in fact they are old."[28]

Repairing offered many town dwellers the chance to earn a livelihood, since only a minor portion of repair work—especially textile repair—seems to have been undertaken within the household. Although there are a number of examples of servants who mended clothing,[29] and of efforts by several institutions to teach both girls and boys sewing skills,[30] the fabric was valuable and could be easily ruined by unprofessional attempts at mending.[31] Thus, more complex tasks that required specific skills and equipment were usually given to specialized external laborers, such as tailors or seamstresses. In some towns, especially larger centers dominated by craft and trades, guilds for repair workers were even created: Nuremberg, for example, had 117 master cobblers and only 110 master shoemakers in 1785.[32] These formal differentiations seem limited to the larger urban crafts, such as tailors and shoemakers, but they sometimes also appear among the metalworkers, where tinkers or locksmiths could form their own occupational groups.[33]

In addition, many craftsmen producing specific goods also engaged in repairing them. Especially for poorer master craftsmen, who worked on their own or were unable to bear the cost of the materials required for production, repair work formed an important source of income.[34] The relevance of repairing for urban craftsmen is underlined by frequent legal disputes about what activities were permitted for different crafts.[35] These frictions, which obviously represent only a small part of the actual disagreements, provide rare sources of information about patterns of repair within urban craft. Even minor repair work could lead to protests by guild members: In 1775, for example, the Viennese locksmiths' guild complained about a dealer in old iron who repaired wooden vices with metal fittings that a customer had bought.[36] When a Viennese cobbler tried to gain permission to sell shoes that had been repaired with "new" leather in the early 1780s, the shoemakers' guild opposed it, as cobblers were only entitled to repair and sell used shoes.[37]

In many master craftsmen's workshops, repairs were frequently carried out by apprentices or younger journeymen—that is, by less-qualified workers who could not be fully utilized in the production process. But in some crafts, experienced journeymen were obliged to do repair work as well, while in other cases such work could be given as a reward.[38] In addition, women and children were often used for preparatory work.[39]

Along with craftsmen within guilds, many other actors engaged in repair work. Urban authorities (magistrates or feudal lords) could issue formal permissions that often bound workers to specific places or products. In Vienna, such permits were issued especially from the early eighteenth century onward and they included, among countless others, tailors, cobblers, and locksmiths.[40] But many repair workers acted without formal permits. Usually these informal actors were labeled (by guildsmen) as *Störer* or *Pfuscher* (bumbler), implying their dubious qualifications. For the most part, these *Störer* were journeymen who had received formal training and worked outside masters' households.[41] Considerable numbers of repair workers seem to have been members of the town guard and the city militia; although they mostly undertook maintenance work for the military, they also had civilian customers. Often these activities were tolerated by the military authorities (who were responsible for these soldiers and for "military" areas) due to the soldiers and their dependents' uncertain living conditions; sometimes military officials even issued formal authorizations. In Vienna, members of the *Stadtguardia*, the town guard, which existed from the sixteenth century until the 1740s,[42] acted as tinkers and cobblers and mended clothing.[43] When nine former members of the guard tried to obtain permission to continue selling repaired shoes in a Viennese suburb in 1760, they even referred to themselves as "by profession menders of old shoes."[44]

The practices of repairing and secondhand trading were closely aligned, since used goods were frequently repaired or altered to be resalable or to achieve higher prices. Often the traders themselves undertook the repairs, benefitting from the circumstance that many of them had been trained as apprentices or even journeymen (see below). Interestingly, in eighteenth-century Vienna, secondhand dealers were not allowed to repair commodities themselves; instead they had to engage master craftsmen. The fact that this formal rule was not obeyed in practice led to recurring conflicts.[45] Repairing seems to be specifically relevant for traders specializing in textiles or metal wares: in the late eighteenth century, several Viennese secondhand dealers even tried to obtain specific permission to repair or rework goods such as old metal and furniture.[46] Another option was to use craftsmen from inside the guild system for repairs—as the rules demanded—or to subcontract the (obviously cheaper) clandestine workers such as "military" craftsmen or *Störer*.[47] Repair workers could also be found in urban secondhand marketplaces, which existed in many cities until the mid-nineteenth century. The Viennese *Tandelmärkte* offered—in addition to a wide variety of secondhand goods and a selection of newly manufactured products—the mending of shoes and the repairing or reworking of metal wares *in situ* (see figure 7.1).[48]

There were also other significant ties between new and old: sometimes repair workers seem to have been engaged in (small-scale) trading with used or

Figure 7.1. Licenses on the Viennese *Tandelmarkt*, 1772–1791 (N = 6026)

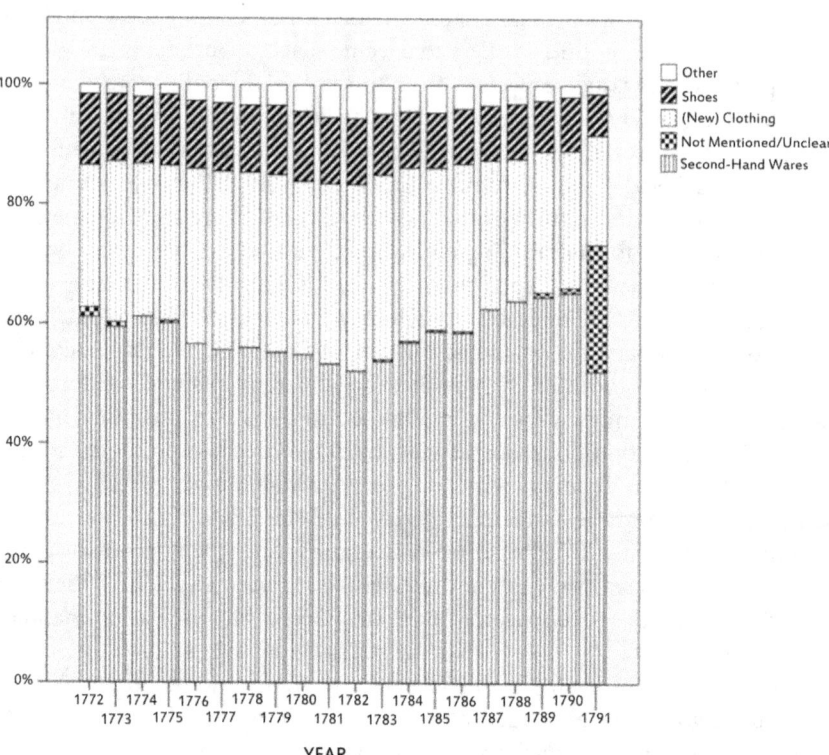

The heading *Shoes* implies the repair and/or trade in shoes; *Others* includes various occupations (such as locksmiths or petty traders); *(New) Clothing* refers to the making, trading, or repairing of clothing. The 1791 cadaster mentions neither fields of employment nor the traded products, but these could be partially reconstructed using preceding years' data.

Source: WStLA, Steueramt, B10/7-26 (Viennese taxation cadasters, 1772 to 1791).

newly manufactured goods. This affected mostly itinerant actors (see below), but other workers, too, sometimes offered to trade used goods for repaired ones.[49] Moreover, the process of repairing could involve the partial use of new materials.[50]

Some repair workers were itinerant and operated on the streets or in their customers' homes. Itinerant actors could especially be found in rural areas, and sometimes they combined repair work with peddling or collecting scrap materials.[51] A former day laborer from Salzburg, for example, sold used shoes that his spouse had repaired in the suburbs.[52] Often repairs were undertaken outside shops or the workers' dwellings, on the streets, in squares, or at the town gates, as a number of archival documents and pictorial and literary sources indicate.⁵³ The artifacts to be repaired could be left by the customer or picked up by the worker himself; repair work—especially smaller repairs—could also be undertaken while the customer waited.⁵⁴ A Viennese publication of the 1840s mentions tailors who would mend clothing in the stables of big inns for porters, coachmen, and servants; others would offer their services to travelers lodging in these inns and other accommodations. The same publication notes that tailors who mended in private houses would pay small sums to domestics in order to gain access to the households.⁵⁵

Repairing obviously served as an economic niche for informal actors who were excluded from regular occupations, and therefore it included particularly large numbers of ethnic or religious minorities. Urban Jews frequently participated in secondary markets such as pawnbroking, peddling, and secondhand trade, as well as repairing—Jewish actors mainly seem to have mended clothing.⁵⁶ We do not know much about these "economies of makeshift," since systematic empirical studies are still lacking. But there are several clues—in contemporary literature as well as in earlier historiographical studies—that suggest the relevance of the repair trade for the Jewish laboring poor. In the early eighteenth century, the Italian physician Bernardino Ramazzini included in his investigation of artisans' diseases (*De morbis artificum*) a chapter on the health impairments of Jews (*De morbis Judaeorum*), obviously influenced by his observations in northern Italy. Ramazzini noted in passing that Jews—and especially Jewish women—often undertook repair work, mainly on clothing, but also shoes.⁵⁷ Another example of Jewish actors in the repair trade is documented in the Moravian town of Proßnitz: In 1820, Lazar Schreiber, who specialized in trading used military uniforms, bought 168 pounds of used and cut-up military jackets along with other goods from a military institution in Lower Austria; he planned to rework these jackets into coats for the rural population.⁵⁸ The substantial size of this transaction implies, on the one hand, extensive structures of reprocessing and redistribution, and on the other hand, the existence of a market for clothing that was reworked in this way.

The lack of academic knowledge about the everyday economic practices of outcasts during the premodern period is also true for the Romani people.⁵⁹ Here, as in the case of Jewish actors, we have several indications that repairing was an important option for earning a living. Newer studies mention the mending of metal wares, shoes, and umbrellas, which could be specializations or part of multiple occupational strategies.⁶⁰ The *Rom* in southeastern Europe often worked as smiths, which is an occupation closely linked to repair work.⁶¹ Today, *Kalderásch* is one of the names used by some Romani groups to refer to themselves—a term similar to the Romanian word for tinker (*căldărar*).⁶² Literature such as Heinrich-Moritz-Gottlieb Grellmann's tendentious *Versuch über die Zigeuner* (Essay on the Gypsies), published in 1783, also mentions these fields of repair, which often appear to have been performed by itinerant workers.⁶³ In addition to repression and persecution, Romani people were subjected to exclusion from many of the same fields of employment as Jews. The strategies of both groups seem to resemble each other: they became active in economic niches, which could include repair work. Astonishingly, a number of vagrant groups seem to have taken up similar activities quite independently—for example, the Dutch *Reizigers* (travelers), who peddled and repaired, or the *Tinkers,* who offered their services in Ireland and Scotland until the twentieth century.⁶⁴

As sources such as contemporary literature and engravings indicate, as well as tax records and the prices that can be calculated for repair work, actors who specialized in repairing were usually part of the urban laboring poor. Prices for repair work are rarely mentioned explicitly. The information from Salzburg presented earlier provides neither the amount of work involved for these repairs nor the possible costs for additional material; thus the sums (12 kr. for mending a jacket or 28 kr. for repairing a pair of shoes) can only be evaluated in comparison to a day laborer's wage, which in the late 1780s was 11 kr.⁶⁵ The taxes or tax classifications for repair workers fit this picture: the mid-nineteenth-century Bavarian and Austrian taxation systems both put repair workers such as tinkers, cobblers, and *Flickschneider* (tailors who mended clothing) in the lowest tax class.⁶⁶ In the late eighteenth century, cobblers on the Viennese secondhand markets usually paid 2 fl. per year for their licenses, which is rather low even compared to other traders and craftsmen at these markets (see figure 7.2). This sum was equivalent to the so-called *Mitleidigensteuer,* which had to be paid by poor town dwellers if they wished to remain burghers.⁶⁷

Laborers doing repair work were exposed to similar workloads as their colleagues in the production process: According to the German physician Georg Adelmann, who published his observations in the early nineteenth century, shoemakers and cobblers often suffered from physical deformities and damages to their internal organs due to their bent posture while working. The frequent occurrence of eye diseases could be interpreted as a consequence of working with little or artificial light.⁶⁸ Ramazzini observed similar health risks for the laborers reworking used textiles: Lack of light caused vision problems while the posture and working at open windows had serious health consequences. In addition, working with used clothing bore the risk of catching infectious diseases.⁶⁹

Because actors engaged in repair work were poorly paid and often part of the lower social classes, there were followed by negative stereotypes. Poverty dominates portrayals of that period,⁷⁰ and repair workers were often met with assumptions of fraud and criminal behavior, especially if they were itinerant actors or members of ethnic or religious minorities.⁷¹ Several accusations claim that repaired products were sold as "new" ones.⁷² Altogether this picture seems biased, and the portrayals—especially those of the late eighteenth and early

Figure 7.2. Mean annual taxes of cobblers compared to other traders and craftsmen on the Viennese *Tandelmarkt* in florins, 1772–1791 (N = 6026)

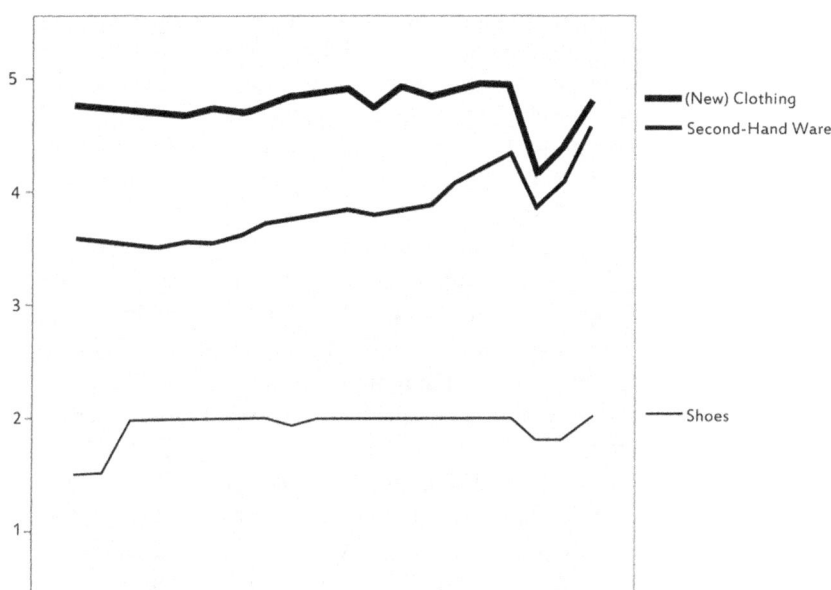

See figure 7.1 for notes.
Source: WStLA, Steueramt, B10/7-26 (Viennese taxation cadasters, 1772 to 1791).

nineteenth centuries—reflect bourgeois attitudes toward poverty and social outcasts, viewing them from a moralistic, romantic, or exotic perspective. In contrast, repairs and repair workers in their everyday contexts are hardly addressed. Perhaps this is a consequence of repairing being an omnipresent phenomenon. Premodern people dealt with repair workers on a regular basis, and probably with a certain amount of trust thanks to the high value of many commodities.

Secondhand: Consumption and Trade

In premodern society, the reuse and exchange of used goods was part of everyday life. Objects were valuable and circulated as presents, nonmonetary payments, bequests and loans, or as loot, and they were traded both privately and publicly.[73] The usefulness of secondhand trade for the household economy and daily consumption was multiple: Firstly, used goods could be purchased at significantly lower prices than new commodities. It is difficult to compare prices due to a lack of sources, and when prices are mentioned, we usually do not know the condition, appearance, or quality of the particular item. Nevertheless, as an example I have tried to calculate prices for men's clothing in Salzburg at the end of the eighteenth century. While by no means perfect, it shows the differences between "first" and "secondhand" prices for clothing, even for cheap apparel (see figure 7.3).

When we compare, we see a broad range of prices for secondhand goods, which obviously depend on their quality or condition. Interestingly, however, a significant amount of money could be saved by purchasing secondhand: the difference between the categories "new" and "used" amounts to more than 20 fl. (for the cheapest) and 14 fl. (for the more expensive option), which would be equivalent to what a day laborer in Salzburg in the late 1780s earned in 110 or 80 days, respectively.[74]

Secondhand trade also helped to convert unnecessary or dispensable belongings and commodities from bequests or insolvencies into cash. Considering the aforementioned sums—even the 3 fl. 30 kr. for a jacket of "poor quality" would have apparently cost a mason in Salzburg the equivalent of nine days' work[75]—one can understand why premodern objects were well treasured by their owners, especially by those who were not so well off. Secondhand trade was not just a necessity, it also created new economic possibilities. By investing in material goods that only slowly diminished in value and were easily resold, people avoided the problem of inflation. In a "cash-starved economy," objects could function as "alternate currencies" used as barter (even in exchange for newly manufactured commodities) or as pledges, for example in taverns.[76] Secondhand markets also seem to have facilitated spatial mobility: belong-

Figure 7.3. Prices of Men's Clothing in Salzburg, 1770–1790

	newly manufactured[a]	secondhand
shirt	2 fl. 24 kr.	1 fl. 32 kr.[b]
jacket	15 fl.	3 fl. 30 kr. ("poorer quality")–7 fl. ("fine")[c]
trousers	5 fl.	30 kr.[d]–2 fl. 3 kr.[e]
pair of shoes	1 fl. 36 kr.	24 kr.[e]
total	24 fl.	3 fl. 41 kr.–9 fl. 36 kr.

Notes: (a) "for a male farm laborer" (*für einen Bauernknecht*—fictive calculation, 1798); (b) auction (1777); (c) estimate (1777); (d) transaction (1777); (e) transactions (1784).

Source: Stöger, *Märkte,* 211.

ings, especially bulky objects such as furniture and household goods, could be sold before traveling and a substitute could be purchased (or rented) on arrival at the final destination. In addition, by purchasing (cheaper) secondhand items (especially clothing), people could imitate the appearance of members of higher social strata, or dress more respectably or fashionably than their funds would otherwise allow. The consumption of used goods was very flexible and was a basic element of everyday economy, especially for members of lower urban strata. Presumably, middle and higher social strata were less involved in the general secondhand market. Obviously, specific market segments existed for wealthier customers, especially for precious or rare goods. Nevertheless, the processes of passing used commodities down the social hierarchy seem to have been significant.[77]

Secondhand markets offered a broad range of prices and appearances, and the products were immediately available: clothing was particularly common due to its relatively high value, but furniture and other everyday objects and even scrap materials were sold as well. Illegally acquired objects (e.g., stolen or plundered goods) also formed a significant part of the secondhand circulation.[78] The markets and traders were closely connected both personally and spatially to numerous other aspects of the urban economy: to handicraft workers (by repair work and the supply of materials), but also to pawnbrokers and traders of new products. The Viennese *Tandelmärkte* had many secondhand dealers who sold unused goods, often against the rules.[79] Urban secondhand trade tended to be limited to a local market, but there is also evidence of transregional transfers of used textiles, for example rags being sold for paper production, which increased despite steady interdictions during the eighteenth century.[80] In the previous century, worn clothing was already being commercially exchanged beyond national borders, for example between France and Britain. Starting in the late eighteenth century, used textiles were exported from several important English ports (for example, Bristol) overseas to Australia, as well as to African and Asian colonies.[81]

The number of professional (formal and sworn or privileged) traders who can be traced in many cities from the late Middle Ages onward might underline the importance of dealing with secondhand goods within the urban economy. Paris and Vienna can serve as examples, though the following estimates have to be considered low, since they do not include all licensed or informal people involved. Around 1725, about 700 specialized retailers officially sold secondhand goods in Paris, which had a population of around 500,000.[82] By the mid-eighteenth century, probably 400 licensed traders were operating in Vienna (population ca. 175,000),[83] and by the second decade of the nineteenth century, the number of dealers taxed by the Viennese municipality had grown to over 500 (see figure 7.4).

Figure 7.4. Secondhand Traders Taxed by the Viennese Magistracies, 1738–1803

Year	Inner City Dealers	Suburban Dealers	Licensed Dealers	Licenses for the *Tandelmarkt* (Only Secondhand)	Sum	Per 1000 Inhabitants
1736	18 (9.6 percent)*	24 (12.8 percent)*	?	146 (77.7 percent)*	188	1.18
1742	18 (7.5 percent)	60 (24.9 percent)	?	163 (67.6 percent)*	241	1.5
1754	19 (7.1 percent)	85 (31.6 percent)	?	165 (61.3 percent)*	269	1.53
1764	18 (71 percent)	77 (30.2 percent)	?	160 (62.7 percent)*	255	1.42
1771	18 (7.2 percent)	66 (26.4 percent)	12 (4.8 percent)	154 (61.6 percent)	250	1.3
1781	18 (7.9 percent)	55 (24.1 percent)	10 (4.4 percent)	145 (63.6 percent)	228	1.11
1791	17 (4.6 percent)	92 (24.8 percent)	9 (2.4 percent)	253 (68.2 percent)*	371	1.72
1803	18 (4.1 percent)	95 (21.7 percent)	6 (1.4 percent)	319 (72.8 percent)*	438	1.89
1812	18 (3.4 percent)	97 (18.1 percent)	45 (8.4 percent)	376 (70.1 percent)*	536	2.25

*Estimate

Source: Stöger, Märkte, 157.

Secondhand wares were traded in shops and stalls or on the street and in public places. Larger cities such as Paris, Amsterdam, Nuremberg, Prague, and Vienna even had their own secondhand special markets held either daily or several times a week. In addition, secondhand goods were disseminated by auctions, which were held in urban areas almost on a daily basis and offered forfeited pledges from pawn shops or pawnbrokers or private belongings.[84] Used commodities could also be traded at seasonal markets and fairs. Furthermore, a significant number of secondhand traders acted as peddlers. Peddling was an option for poorer traders, such as soldiers and their dependents, who often had no formal authorization. These traders also served as distributors or purchasers for other (formal and/or stationary) retailers.[85] Premodern itinerant traders had an important function for supplying rural areas, as several studies have already emphasized.[86] Mostly, these rural peddlers seem to have sold new (petty) goods, but there are indications of offering secondhand ware or bartering used goods and scrap materials for other commodities.[87]

In urban Europe, the process of restricting secondhand trade to licensed actors varied significantly according to place and time: some cities did not issue formal licenses for such traders until the nineteenth century, while others, such as Vienna and Antwerp, introduced privileges that could be administered by corporations. But these guild actors usually remained a minority and served specific (often upper-class or highly specialized) market segments.[88] In German-speaking territories, surveillance and formalization efforts, which were often linked to the authorities' mistrust or taxation attempts, seem to have intensified from the late eighteenth century onward. Unsurprisingly, due to the omnipresence of secondhand transactions, informal engagements were widespread. Although impossible to quantify, these formed a significant part of the overall secondhand trade.[89]

The commercial exchange of used goods provided an important source of income for many town dwellers and a temporary (supplementary and occasional), or middle or long-term, occupation for poor urban laborers. Moreover, the secondhand trade often offered women and members of ethnic or religious minorities the possibility to earn a living. It was mainly nonelite members of the urban population who acted as secondhand dealers: domestic servants, casual workers, or soldiers (and their dependents). Many journeymen (especially tailors and shoemakers) could turn to secondhand trade, benefiting from their specialized knowledge. The socioeconomic diversity among licensed secondhand dealers appears to have been significant, ranging from quite respectable and wealthy individuals who owned stores in town and even hired helpers, to retailers who displayed their few and humble goods on the street. However, secondhand trade, and especially its informal side, was undoubtedly dominated by less-affluent members of society who were often considered only marginally respectable. Urban Jews, having been excluded from many occupations, became active on secondary markets as well. For them, activities such as lending money on pawned goods and selling forfeited pledges provided a natural link to trading with used objects; in addition, the secondhand trade, being a rather low and often hardly regulated market, was more accessible to newcomers (which could include migrants).[90]

Trading used goods was often perceived as not especially respectable, and secondhand dealers were frequently seen as contaminated by association with old and unclean goods. At the same time, more discerning appraisals stress the aspects of necessity and utility. Accounts and sources of that period are full of generalized complaints about dubious methods of buying and selling. These refer to the arbitrary assessment of prices, secret arrangements in auctions, and the sale of poor-quality products. Problems like trading with stolen or infected goods also fostered ambivalent or negative perceptions. Such disparaging perceptions increase throughout the eighteenth century, but seem limited to individuals from the middle and upper social classes. Common townsfolk worked with secondhand dealers on a daily basis and had close contact with official handicraft workers engaged in repair work (for trade and because of the artisans' need for scrap materials). The upper classes' prejudices seem to have affected legislation, as legal authorities often shared their distrust of the secondhand trade, especially toward its itinerant forms and the engagement of members of religious or ethnic minorities.[91] A travel account from the early nineteenth century depicting Jewish dealers in London is a rather typical portrayal of petty urban secondhand trade:

> We passed through a great crowd of dirty ragged people, to the number of some hundreds. They appeared to be very busy in displaying and examining the old clothes which they were pulling from bags. This, I was informed, was Rag Fair. It is held here every evening for the sale of old clothes, which are collected all over London, principally by Jews who go about with bags on their shoulders crying, with a peculiarly harsh guttural sound, "clothes, clothes, old clothes." You will meet them in every street and alley in London, and, at evening, they repair to Wapping, where a grand display is made of every species of apparel, in every stage of decay.[92]

Reports of secondhand trade endangering public safety can regularly be found in administrative sources. These perceptions often relate to the selling of stolen or infected goods. Nevertheless, surprisingly insightful observations can be found as well, where magistracies or feudal lords point out the necessity and importance of secondhand trade for the urban laboring poor.[93]

Conclusion

Undoubtedly, the premodern "sustainable" strategies were mainly driven by economic motives, but they also seem rooted within a premodern mentality of thrift. The reuse and repair of objects was common within premodern daily life and could also be found outside the laboring poor. The relatively high prices of many everyday commodities contributed to the omnipresence of repair and secondhand trade and made them important parts of the premodern urban infrastructure. For consumers, reuse and repair were necessities as well as a source of opportunity: Repairing items or purchasing secondhand goods helped to reduce expenses and to relieve tight budgets. Secondhand trade helped to convert belongings into ready cash and facilitate spatial mobility. Repairing and secondhand consumption could help people to dress in a respectable and fashionable way. Furthermore, secondary markets offered significant potential to earn a livelihood; especially for the urban laboring poor and for women and other actors excluded from other occupations.

There were close ties between old and new: producers engaged in repair works, and secondhand consumption cannot be isolated from first-hand consumption, for consumption was often situational and dependent on specific needs and economic possibilities. The perceptions of actors involved in these markets seem ambivalent: some stressed the utility aspects, whereas others, particularly members of the middle and upper social classes, express negative perceptions that focus on dubious or unlawful practices. These prejudices were intensified by the involvement of outcasts in secondary markets and by a principal mistrust of itinerant actors that was partially shared by the authorities.

As yet, little is known about the development of secondary markets during the industrial period. Industrialization and mass production cheapened many everyday commodities (especially clothing), and thereby changed the usage of goods and patterns of daily consumption. We can see both signs of change and persistence: the large urban marketplaces for secondhand commodities disappeared in the second half of the nineteenth century, and the secondhand trade seems to have increasingly lost its relevance for large parts of the urban population. In contrast, repairing does not appear to have been marginalized in the course of the nineteenth century. Due to the expansion of industrial mass production, many craftsmen, especially tailors and shoemakers, were forced to engage in the putting-out system or in repair work. In addition, new fields of repair emerged, mainly in an industrial context. Overall, literary and pictorial sources show that secondary markets persisted—especially among the laboring poor—as forms of consumption and occupation. In the Western world, a fundamental caesura in the usage of goods seems to have occurred after the Second World War with the emergence of the throwaway society—or, as Swiss Historian Christian Pfister labeled it, the "1950s syndrome."[94]

Georg Stöger is a postdoctoral assistant in economic, social, and environmental history at the University of Salzburg. His research currently focuses on changes in urban environments during the eighteenth and nineteenth centuries. Most recently, he has published on premodern and modern practices of material reuse, such as the secondhand trade, repairing, and recycling.

Notes

I would like to thank Luisa Pichler (Salzburg) for editing my paper and giving valuable suggestions.

1. Utz Jeggle, "Vom Umgang mit Sachen," in *Umgang mit Sachen: Zur Kulturgeschichte des Dinggebrauchs. 23. deutscher Volkskundekongreß in Regensburg vom 6.–11. Oktober 1981* (Regensburg, 1983), 11–25; Ludolf Kuchenbuch, "Abfall: Eine stichwortgeschichtliche Erkundung," in *Mensch und Umwelt in der Geschichte*, ed. Jörg Calließ et al. (Pfaffenweiler, 1989), 257–276; Friedrich Jaeger, ed., *Enzyklopädie der Neuzeit* (hereafter EdN), 16 vols. (Stuttgart and Weimar, 2005–2011), vol. 1, 11–13, under "Abfall."
2. See as overviews: Donald Woodward, "'Swords into Ploughshares': Recycling in Pre-Industrial England," *The Economic History Review* 38, no. 2 (1985): 175–191; Ruth

Oldenziel and Heiker Weber, "Recycling Reconsidered," *Contemporary European History* 22, no. 3 (2013): 347–370; Georg Stöger and Reinhold Reith, "Western European Recycling in a Long-Term Perspective: Reconsidering Caesuras and Continuities," *Jahrbuch für Wirtschaftsgeschichte* 56, no. 1 (2015): 267–290.

3. Susan Strasser, *Waste and Want: A Social History of Trash* (New York, 1999), 21–52, quotation on 28.
4. On pictorial evidence of trades dealing with recycling and reusing, see Karen F. Beall, *Kaufrufe und Straßenhändler: Eine Bibliographie* (Hamburg, 1975), 97, 103, 465, 469.
5. For example, in Paul T. Anastas and Julie B. Zimmerman, "Design through the 12 Principles of Green Engineering," *Environmental Science & Technology* 37, no. 5 (2003): 94–101; Herman E. Daly, *Ecological Economics and Sustainable Development: Selected Essays of Herman Daly* (Cheltenham, 2008); see also the introduction to this volume.
6. Christian Pfister, ed., *Das 1950er Syndrom: Der Weg in die Konsumgesellschaft*, 2nd ed. (Bern, 1996), 65.
7. Karl Bücher, *Die Berufe der Stadt Frankfurt am Main im Mittelalter* (Leipzig, 1914), 18.
8. See Reinhold Reith, "Recycling im späten Mittelalter und der frühen Neuzeit: Eine Materialsammlung," *Frühneuzeit-Info* 14 (2003): 47–65, here 47–48.
9. See as recent overviews: EdN, vol. 11, 58–61, under "Reparieren"; ibid., vol. 13, 791–793, under "Trödel"; ibid., vol. 14, 1079–1085, under "Wiederverwertung."
10. Beverly Lemire, *Dress, Culture and Commerce: The English Clothing Trade before the Factory, 1660–1800* (Basingstoke, 1997); Eadem, "Shifting Currency: The Culture and Economy of the Second Hand Trade in England, c. 1600–1850," in *Old Clothes, New Looks: Second Hand Fashion*, ed. Alexandra Palmer and Hazel Clark (Oxford and New York, 2006), 29–47; Laurence Fontaine, ed., *Alternative Exchanges: Second-hand Circulations from the Sixteenth Century to the Present* (New York and Oxford, 2008); Jon Stobart and Ilja Van Damme, eds., *Modernity and the Second-Hand Trade: European Consumption Cultures and Practices, 1700–1900* (Basingstoke, 2010); Georg Stöger, *Sekundäre Märkte? Zum Wiener und Salzburger Gebrauchtwarenhandel im 17. und 18. Jahrhundert* (Vienna and Munich, 2011); Reinhold Reith and Georg Stöger, "Reparieren – oder die Lebensdauer der Gebrauchsgüter," *Technikgeschichte* 79, no. 3 (2012): 173–184; *Flick-Werk: Reparieren und Umnutzen in der Alltagskultur; Begleitheft zur Ausstellung im Württembergischen Landesmuseum Stuttgart vom 15. Oktober bis 15. Dezember 1983* (Stuttgart, 1983).
11. For overviews on the premodern period, see Reinhold Reith, *Umweltgeschichte der Frühen Neuzeit* (Munich, 2011); Joachim Radkau, *Nature and Power: A Global History of the Environment* (Washington, DC, 2008); Martin Knoll and Reinhold Reith, *An Environmental History of the Early Modern Period: Experiments and Perspectives* (Vienna, 2014).
12. Valentin Groebner, *Ökonomie ohne Haus: Zum Wirtschaften armer Leute in Nürnberg am Ende des 15. Jahrhunderts* (Göttingen, 1993), 116, 180.
13. See EdN, vol. 12, 791–793, "Sparsamkeit."
14. Groebner, *Ökonomie*, 220–221; Stöger, *Märkte*, 207.
15. Olwen Hufton, *The Poor of Eighteenth-Century France 1750–1789* (Oxford, 1974), 69–127; see Laurence Fontaine and Jürgen Schlumbohm, "Household Strategies for Survival: An Introduction," *International Review of Social History* 45 (2000): 1–17.

16. Stöger, *Märkte*, 212–213.
17. Reith, "Recycling," 49.
18. Woodward, "Swords," 186.
19. *Salzburger Intelligenzblatt*, 4 August 1798; one *Gulden* (fl.) equals 60 *Kreuzer* (kr.).
20. Eighteen kr. in the 1780s; see Archiv der Stadt Salzburg (Salzburg municipal archive, hereafter AStS), Privatarchiv 1.172, Franz Anton Spängler's accounts, vol. 4 (1772–1785), without pagination (24 March 1781).
21. Státní okresní archiv Prostějov (Prostějov municipal archive), Archiv města Prostějov, Box 4/Fasz. 30 (Stará registratura. Akta hospodař. úřadu etc. 1786).
22. AStS, Privatarchiv 1.172, Franz Anton Spängler's accounts, vol. 3, (1760–1771), without pagination (20 May 1770).
23. Salzburger Landesarchiv (Salzburg district archive, hereafter SLA), Verlaß Stadtsyndikat, No. 610 (bill, 8 October 1789).
24. Database of the account books of Franz Anton Spängler (1733–1785). This database was developed by a research group at the University of Salzburg and contains roughly 21,000 entries on different expenses of the Spängler household.
25. See Leopold Mozart's letter to his son, 8 January 1781 in *Mozart: Briefe und Aufzeichnungen; Gesamtausgabe*, ed. Internationale Stiftung Mozarteum Salzburg (Kassel, 2005), vol. 3, 85; and Leopold Mozart's letters to his daughter, 20 October 1786, in ibid., 597; and, 8 December 1786, in ibid., 618.
26. See *Geschichte des Königlichen Potsdamschen Militärwaisenhauses [...]* (Berlin and Posen, 1824), 311–312; August v. Witzleben, *Grundzüge des Heerwesens und des Infanteriedienstes der Königlich Preußischen Armee*, 2nd ed. (Berlin, 1850), 250.
27. Wolfgang Amadeus Mozart's letter to his father, 27 December 1780, in *Mozart*, vol. 3, 73; SLA, Verlaß Stadtsyndikat, No. 333 (calculation, 6 March to 13 September 1784).
28. "Man flickt, man dreht, man näht, ... man hefft alles untereinander, damit es nur nach der Mode sey... ; das meiste scheinet neu zu seyn, welches doch in der Sache selbsten alt"; Johann Valentin Neiner, *Neu Ausgelegter Curioser Tändel-Marckt der jetzigen Welt in allerhand Waaren und Wahrheiten vorgestellet [...]* (Vienna and Brno, 1734), 240.
29. See Marforius, *Kurtze Beschreibung Des zum theil liederlichen Lebens und Wandels Derer anjetzo in grossen Städten sich befindenden Dienst-Mägde [...]* (s.l. s.a. [probably published in Leipzig around 1717]), 8; SLA, churf. u. k. k. Regierung, XXXVI/X II/No. 23 (petition Joseph Treiber, 29 November 1787).
30. These seem to have intensified in the second half of the eighteenth century; see the examples in Gernot Heiß, "Erziehung der Waisen zur Manufakturarbeit: Pädagogische Zielvorstellungen und ökonomische Interessen der mariatheresianischen Verwaltung," *Mitteilungen des Instituts für Österreichische Geschichtsforschung* 85 (1977): 316–331, here 323; and Reinhold Vormbaum, ed., *Die evangelischen Schulordnungen des achtzehnten Jahrhunderts* (Gütersloh, 1864), vol. 3, 45.
31. John Styles, *The Dress of the People: Everyday Fashion in Eighteenth-Century England* (New Haven, CT, 2007), 73–75; Judith G. Coffin, "Gender and the Guild Order: The Garment Trades in Eighteenth-Century Paris," *The Journal of Economic History* 54 (1994): 768–793, here 771.
32. Reith, "Recycling," 50–51.
33. Friedrich Hornschuch, *Aufbau und Geschichte der interterritorialen Kesslerkreise in Deutschland* (Stuttgart, 1930), 99–100; EdN, vol. 11, 783–785, under "Schlosser."

34. For the depiction of cobblers in an 1840s Viennese play, see Johann Baptist Moser, *Das Wiener Volksleben in komischen Scenen mit eingelegten Liedern [...]* (Vienna, 1842), vol. 2.
35. Jutta Zander-Seidel, *Textiler Hausrat: Kleidung und Haustextilien in Nürnberg von 1500–1650* (Munich, 1990), 287, 378–379.
36. Wiener Stadt- und Landesarchiv (hereafter WStLA), Alte Registratur, A1 7/1755.
37. WStLA, Alte Registratur, A2 763/1782 (report of the Viennese magistracies, 30 November 1782).
38. Reinhold Reith, *Lohn und Leistung: Lohnformen im Gewerbe 1450–1900* (Stuttgart, 1999), 190, 293–294.
39. H. E. Schmiederer, ed., *Mein Lebensmorgen: Nachgelassene Schrift von Wilhelm Harnisch; Zur Geschichte der Jahre 1787–1822* (Berlin, 1865), 20, 23; August Daniel v. Binzer, *Venedig im Jahre 1844* (Budapest, 1845), 170.
40. Margit Altfahrt, "'Den Professionisten ist wider ihre Störer alle Assistenz zu leisten...' Unbefugte Schneider im Wien des späten 18. Jahrhunderts," *Jahrbuch des Vereins für Geschichte der Stadt Wien* 52/53 (1996/1997): 9–32, here 19; Stöger, *Märkte*, 172–173.
41. EdN, vol. 12, 1050–1052, under "Störer."
42. On the Viennese "Stadtguardia," see Stöger, *Märkte*, 91–92.
43. WStLA, Alte Registratur, A2 348/1766; ibid., A1 70/1712; ibid., A2 235/1767; see Johann Basilius Küchelbecker, *Allerneueste Nachricht vom Römisch-Kayserlichen Hof [...]* (Hannover, 1730), 447.
44. "der Profession alte Schuhflicker"; Österreichisches Staatsarchiv (Austrian state archive, Vienna), Kriegsarchiv, Hofkriegsrat, Protokoll in Publicis 1760, fol. 821a.
45. WStLA, Alte Registratur, A2 43/1774; see Stöger, *Märkte*, 40–41, 117–118.
46. This request was rejected by the Viennese authorities since it concerned the privileges of several craftsmen. See WStLA, Alte Registratur, A2 452/1781 (Viennese magistracies report, undated [before 5 September 1781]).
47. Stöger, *Märkte*, 116–117.
48. Ibid., 38–40.
49. Binzer, *Venedig*, 170; Christoph Weigel, *Abbildung Der Gemein-Nützlichen Haupt-Stände [...]* (Regensburg, 1698), 651.
50. WStLA, Alte Registratur, A2 763/1782.
51. Chris Glass, "Reparierendes Handwerk," in *Flick-Werk*, 35–42, here 35; Wolfgang Scheffknecht, "Fremde Wanderkrämer und Keßler in der Grafschaft Hohenems und im Reichshof Lustenau," in *Minderheiten, Obrigkeit und Gesellschaft in der Frühen Neuzeit: Integrations- und Abgrenzungsprozesse im süddeutschen Raum*, ed. Mark Häberlein and Martin Zürn (St. Katharinen, 2001), 233–267, here 237–239 and 254–256; Hornschuch, *Aufbau*, 99–100.
52. SLA, churf. u. k. k. Regierung, XXXVI/X II/No. 4.
53. WStLA, Alte Registratur, A2 235/1767; Sean Shesgreen, *Images of the Outcast: The Urban Poor in the Cries of London* (New Brunswick, NJ, 2002), 14–15, 52; Moser, *Volksleben*, IV.
54. See depictions from the first half of the nineteenth century: Moser, *Volksleben*, VI–VII; Friedrich Fabini, *Reise in Italien und zur See nach Spanien* (Hermannstadt, 1848), 9; Wilhelm Hausmann, "Aus dem Leben der Zigeuner in Siebenbürgen," *Oesterreichische Revue. Fünfter Jahrgang. 7. Heft* (Vienna, 1867), 153–163, here 157–158.

55. Moser, *Volksleben*, VII.
56. On the engagement of Jews in secondhand trade and pawnbroking, see Stöger, *Märkte*, 175–180, 221–224.
57. Bernardino Ramazzini, *Untersuchung von den Kranckheiten der Künstler und Handwerker [...]* (Leipzig, 1705), 307–308.
58. Bernhard Heilig, "Die Vorläufer der mährischen Konfektionsindustrie in ihrem Kampf mit den Zünften," *Jahrbuch der Gesellschaft für Geschichte der Juden in der Čechoslovakischen Republik* 3 (1931): 307–448, here 414–415.
59. I use the term *Romani people* for mainly southeastern European *Rom* and western and central European *Sinti*.
60. Thomas Fricke, *Zigeuner im Zeitalter des Absolutismus: Bilanz einer einseitigen Überlieferung; Eine sozialgeschichtliche Untersuchung anhand süddeutscher Quellen* (Pfaffenweiler, 1996), 425–426, 461–462; Angelika Albrecht, *Zigeuner in Altbayern 1871–1914: Eine sozial-, wirtschafts- und verwaltungsgeschichtliche Untersuchung der bayerischen Zigeunerpolitik* (Munich, 2002), 148–149, 155–156.
61. Johann Heinrich Schwicker, *Die Zigeuner in Ungarn und Siebenbürgen* (Vienna, 1883), 121–122.
62. Romani Linguistics and Romani Language Projects, History of the Romani Language, http://romani.humanities.manchester.ac.uk/whatis/language/names.shtml (accessed 12 September 2014).
63. Heinrich-Moritz-Gottlieb Grellmann, *Historischer Versuch über die Zigeuner, betreffend die Lebensart und Verfassung, Sitten und Schicksale dieses Volks seit seiner Erscheinung in Europa, und dessen Ursprung*, 2nd ed. (Göttingen, 1787), 82–84; Hausmann, "Leben", 157–158.
64. Michael H. Faber, "Nichtzigeunerische Landfahrer in Deutschland und anderen europäischen Ländern," in *Zigeuner: Roma, Sinti, Gitanos, Gypsies zwischen Verfolgung und Romantisierung*, ed. Rüdiger Vossen (Frankfurt, 1983), 187–203, here 194 and 198–199.
65. AStS, Pezolt-Akten 297 (calculations on expenses for a Salzburg seasonal market, undated [after 1798]).
66. *Gesetz vom 1. Juli 1856, die Gewerbsteuer betreffend. [...]* (Erlangen, 1859), 181, 187, 191 (for Bavaria); Gustav Höfken, *Die Reform der direckten Besteuerung in Oesterreich auf Grund der Anträge des k. k. Finanzministeriums* (Vienna, 1860), without pagination (11th section) (for the Austrian territories).
67. WStLA, Steueramt, B10/16 (Viennese taxation cadastre 1781), without pagination (rubric "Von denen Mitleydigern Burgern"); ibid., B10/38 (Viennese taxation cadaster 1803), fol. 252b.
68. *Georg Adelmann ausübender Arzt in Würzburg über die Krankheiten der Künstler und Handwerker [...]* (Würzburg, 1803), 79–80, 114–119, 122.
69. Ramazzini, *Untersuchung*, 308–309.
70. As in engravings and literature of the period, e.g.: Shesgreen, *Images*, 14–15, 52; Moser, *Volksleben*; Henry Mayhew, *London Labour and the London Poor* (New York, 1968), vol. 2.
71. Grellmann, *Versuch*, 112–113; Ramazzini, *Untersuchung*, 308. On the perception of ambulant actors on secondary markets, see Stöger, *Märkte*, 62–63.
72. WStLA, Alte Registratur, A2 452/1781.

73. See recent overviews: Laurence Fontaine, "Die Zirkulation des Gebrauchten im vorindustriellen Europa," *Jahrbuch für Wirtschaftsgeschichte* 2004/2: 83–96; Fontaine, *Exchanges*; Stobart and Van Damme, *Modernity*; Stöger, *Märkte*.
74. Eleven kr. per day in 1789. See AStS, Pezolt-Akten 297 (calculated expenses for a Salzburg seasonal market, undated [after 1798]).
75. AStS, Privatarchiv 1.172, Franz Anton Spängler's accounts, vol. 4 (1772–1785), without pagination (24 March 1781).
76. Lemire, *Dress*, 114–115; John Styles, "Clothing the North: The Supply of Non-élite Clothing in the Eighteenth Century North of England," *Textile History* 25, no. 2 (1994): 139–166; Patricia Allerston, "The Market in Second-hand Clothes and Furnishings in Venice, c. 1500–c. 1650," PhD dissertation, University of Florence, 1996.
77. See Stöger, *Märkte*, 207–217.
78. Beverly Lemire, "The Theft of Clothes and Popular Consumerism in Early Modern England," *Journal of Social History* 24 (1990): 255–276; Elizabeth C. Sanderson, "Nearly New: The Second-hand Clothing Trade in Eighteenth Century Edinburgh," *Costume* 31 (1997): 38–48, here 41–45; Brian Sandberg, "'The Magazine of All Their Pillaging': Armies as Sites of Second-hand Exchange during French Wars of Religion," in Fontaine, *Exchanges*, 76–96.
79. Stöger, *Märkte*, 26–32, 37–42.
80. EdN, vol. 14, 1079–1085, "Wiederverwertung."
81. Woodward, "Swords," 179; Bianca M. du Mortier, "Introduction into the Used-Clothing Market in the Netherlands," in *Per una Storia della Moda Pronta: Problemi e ricerche; Atti del V Convegno Internazionale del CISST, Milano, 26–28 febbraio 1990* (Florence, 1991), 117–125, here 123; Sarah Levitt, "Bristol Clothing Trades and Exports in the Georgian Period," in *Per una Storia*, 29–41, here 29, 35–36.
82. Fontaine, "Zirkulation," 87.
83. See WStLA, Alte Registratur, A1 118/1742.
84. Ilja Van Damme and Reinoud Vermoesen, "Second-Hand Consumption as a Way of Life: Public Auctions in the Surroundings of Aalst in the Late Eighteenth Century," *Continuity and Change* 24, no. 2 (2009): 275–305; Melanie Tebbutt, *Making Ends Meet: Pawnbroking and Working-Class Credit* (London, 1984).
85. Stöger, *Märkte*, 49–63.
86. Laurence Fontaine, *History of Pedlars in Europe* (Cambridge, 1996); Wilfried Reininghaus, ed., *Wanderhandel in Europa: Beiträge zur wissenschaftlichen Tagung in Ibbenbüren, Mettingen, Recke und Hopsten vom 9.–11. Oktober 1992* (Dortmund, 1993).
87. Stöger, *Märkte*, 57–58.
88. Ibid., 201–202; Harald Deceulaer, "Second-hand Dealers in the Early Modern Low Countries: Institutions, Markets and Practices," in Fontaine, *Exchanges*, 13–42, here 15.
89. Georg Stöger, "Disorderly Practices in the Early Modern Urban Second-Hand Trade (Sixteenth to Early Nineteenth Centuries)," in *Shadow Economies and Irregular Work in Urban Europe: 16th to early 20th Centuries*, ed. Thomas Buchner and Philip Hoffmann-Rehnitz (Berlin et al., 2011), 141–163.
90. Lemire, *Dress*, 75–79, 118; Stöger, *Märkte*, 169–204, 221–225; Encyclopaedia Judaica (Jerusalem, 1971), vol. 14, 1085–1086, under "secondhand goods and old clothes, trade in"; Betty Naggar, "Old-clothes Men: 18th and 19th Centuries," *Jewish Historical Studies: Transactions of the Jewish Historical Society of England* 31 (1990): 171–191.

91. Stöger, *Märkte*, 233–244.
92. Quoted from Jay Rumney, "18th Century English Jewry through Foreign Eyes 1730–1830," *The Jewish Historical Society of England – Transactions* 13 (1932–35), 323–342, here 335–336.
93. E.g., in WStLA, Alte Registratur, A1 118/1742 (Viennese magistracies report, 20 February 1742); AStS, Zunftarchiv 568 (Salzburg town court report, undated [August 1770]).
94. See Pfister, *1950er Syndrom,* and more recently, Pfister, "The '1950s Syndrome' and the Transition from a Slow-Going to a Rapid Loss of Global Sustainability," in *The Turning Points of Environmental History,* ed. Frank Uekötter (Pittsburgh, PA, 2010), 90–118.

 CHAPTER 8

Waste to Assets
How Household Waste Recycling Evolved in West Germany

Roman Köster

In a strict sense, the history of recycling goes back to the beginning of humankind. For a very long time, the incentive to recycle has been the scarcity of resources, which gave waste materials a value and, eventually, a price. The amount of recycling at any given time was thus an indicator of scarcity and, applied to society, poverty. This implies that the appearance and determinants of recycling have changed considerably in the past decades. Today, the scarcity of plastic is obviously not the reason to recycle it. The recycling of glass is profitable, but it would be easier and cheaper to substitute glass with plastic. The traditional scarcity argument is probably more valid for waste paper and scrap metals, but all in all, we can confirm that recycling today is very different from fifty years ago. Recycling is no longer an indicator of poverty, but of the consequences and challenges of affluence.

This chapter aims to trace the structural changes in recycling in West Germany after the Second World War. This case is especially interesting because since the late 1960s, West Germany, alongside countries like Denmark or Switzerland, became the forerunner and role model of "modern" recycling in industrialized societies.[1] I will explain this by outlining the particular historical circumstances Germany had to deal with: a juxtaposition of massive problems of waste disposal, an ongoing steep growth in the amount of waste, rising prices for energy since the 1970s despite cheap energy being one of the main reasons for Germany's "economic miracle" after the Second World War, and an increasing awareness of environmental protection. These factors went hand in hand with the emergence of a new form of recycling, which most industrialized countries have adopted since the 1980s.

Despite being a "role model," however, I also want to make clear that in the German case, modern recycling was also mainly a way to deal with a growing waste stream and different kinds of scarcity: the scarcity of landfill space, energy, or raw materials. "Sustainability" in this context, therefore, mainly implies attempts to find ways to deal with the consequences of modern consumer societies. Modern recycling is, at least from my point of view, far from representative of a more natural or "self-sustaining" model of production and consumption, although it probably makes modern consumer societies more sustainable.

The history of waste management in Germany during the twentieth century is, despite some new publications, still a comparably new field of research.[2] This is especially true of developments since the late 1960s, with the emergence and institutionalization of a recycling infrastructure, as these were not covered by older literature that mainly dealt with questions of city hygiene or the cultural perception of waste.[3] For this reason, the following chapter is based mostly on archival sources and contemporary technical literature on the problems of waste and recycling.

The End of Old-Fashioned Recycling in West Germany

To speak of "recycling" is to identify things that are "waste" as useless. Indeed, this distinction has proven to be not very clear-cut in recent times: most things that have lost their original utility can be used for something else, such as food waste as compost in the garden, scrap metals for manufacturing, and so on, demonstrated by the long tradition of scrap dealing. Recycling, having been part of everyday life, gradually changed with the introduction of a regular waste-collection service in German cities after the late nineteenth century. Collecting scrap on tips became a regular business and cities frequently commissioned people as subcontractors to carry it out. Munich, as an outstanding example, hired a private firm before the First World War to collect the city's waste. This company ran a facility to sort the waste and look out for valuable materials even if these efforts were seldom profitable.[4] All in all, a secondary market for waste materials existed in which the self-employed could earn a living.[5]

Recycling activities expanded enormously in Nazi Germany and especially during the Second World War. As part of the *Vierjahresplan* (four-year plan), the Nazi government installed a vast recycling system. Nazi organizations such as the *Nationalsozialistische Volkswohlfahrt* (NSV) and the Hitler Youth were obliged to collect waste paper, food waste, and textiles. In the course of this development, the traditional salvage trade, once dominated by Jews, was pushed out of the market—going hand in hand with Nazi racial policy.[6] Besides highly symbolic acts like the melting down of church bells, the state was looking for all sorts of raw materials that could be used for armament. Metals were of special importance, and for this reason the state initiated numerous *Sammelaktionen* (collections).[7]

In the early 1950s, recycling still played a role in waste collection. Scrap dealers were a common feature of street life and children could earn a little money by selling waste paper.[8] But when the German economy experienced the so-called Economic Miracle, people stopped collecting scraps. From the mid-1950s until the early 1970s, there was merely a rudimentary recycling of waste in private households. To explain this, we can turn to psychological factors: recycling was associated with hard times and stopping this practice possibly implied leaving the war years behind. But there are also more "materialistic" and perhaps better explanations. The composition of waste changed: In the early 1950s, waste was typically made up of 30 percent ash, while the rest was scrap metals, textiles, food waste, and paper. Two decades later, packaging, or products without an obvious alternative use (plastic foils for instance), became increasingly important.[9] Furthermore, city structures changed massively during the process of rebuilding West Germany after the Second World War. Outer districts, which had formerly been villages, became real suburbs. This caused waste problems for city centers, as gardens that used to utilize food waste (as compost) ceased to exist.[10] Consequently, there was a lack of opportunities for recycling.

Another possible explanation is housewives' changing behavior. In the 1950s and 1960s, German households experienced a thorough "rationalization" when appliances like vacuum cleaners and washing machines (etc.) were introduced. This brought about a relaxation of housework along with a rising standard of cleanliness, even though the claim that improved cleanliness offsets the time savings achieved by technical improvements was not proven.[11] This rationalization, however, probably did fit well with a rationalized manner of throwing things away: easily and cleanly. It can even be interpreted as practitioners "educating" housewives to transform them into good "wasters." However, we have to bear in mind that those practitioners, who were struggling to cope with a steeply growing waste stream, had no interest in further increasing the amount of waste. Many letters of complaint can be found in the city archives, but the fading of opportunities for recycling was barely mentioned; the complaints mainly dealt with issues of "convenience" or conflicts with neighbors.[12]

The most convincing explanation for the disappearance of recycling, however, is that the economic terms changed. The Nazis had destroyed the traditional infrastructure of the salvage trade, but it is highly probable that others would have stepped in if this business had been lucrative. Due to the rising productivity of manufacturing, falling prices for the manufacture of glass and paper, and generally comparatively low prices for raw materials, the salvage trade lost its economic sense. This was especially true because the productivity of collection and sorting was not growing at the same rate as repair services.[13] Therefore, the economic necessity to collect valuable materials ceased to exist.

This all fits with what historian Christian Pfister has called the "1950s syndrome": a transformation of the relationship between man and nature, which presumably took off in the 1950s and brought about a massive growth in the consumption of energy, goods, countryside, a rising amount of waste, and so on, mainly based on falling prices of fuel and other energy sources.[14] So the recycling of household waste lost its economic sense and, for this reason, stopped.

In Germany's leading journals for waste management, especially *Städtetag* and *Städtehygiene*, the prominent experts agreed that scrap collection no longer made sense economically. On the other hand, they were in favor of some kind of "recycling" on the disposal side. This meant, for instance, waste composting as a solution for the disposal problems in highly agglomerated areas like the Ruhr region. In the end, however, composting turned out to be a solution only for rural regions. Compost extracted from waste in dense industrialized areas proved to be too highly contaminated by heavy metals and other toxic substances. But even without these problems, farmers were generally not willing to buy fertilizer made from rubbish.

Origins of "Modern" Recycling since the Mid-1960s

Despite recycling in households becoming less relevant after the mid-1950s, not all recycling practices ceased. Industrial recycling, for example, remained important. In the early 1970s, West Germany had the highest rate of waste paper recycling in the world (about 30 percent), almost without any private collection.[15] In the case of glass, a traditional system of returning used bottles to the trader persisted, so a beer bottle could be reused about twenty to thirty times. But these systems either depended on a long tradition supported by the manufacturers and distributors (as in the case of the brewing industry) or were only feasible thanks to efficient collection opportunities. Paper recycling would have been successful if large and homogenous amounts of paper could be easily removed from places like firms and big department stores.[16] Even when, like in paper manufacturing, secondary markets existed, the collection costs were a huge barrier to any recycling of household waste.

This rationale strongly suggests that there was no economic justification for recycling in the boom years after the 1950s. Yet the late 1960s saw recycling return to the agenda of practitioners, engineers, the public, and, not least, the state. When the authorities began preparing a general law for the regulation of waste management in the late 1960s (which finally passed the West German Bundestag in 1972), recycling was one of the main strategies to solve the disposal problem. The public started to write letters to cleansing departments, demanding the reuse of valuable materials.[17] How was this possible?

First of all, in the second half of the 1960s, the "economization" of consumer society seemed to climb to a new level. Especially between 1965 and 1970, the production of plastic bottles experienced a steep rise.[18] Within a relatively short period of time, plastics became the second most important packaging material in West Germany (see figures 8.1 and 8.2).

In 1967, the German plastics industry launched an advertising campaign that praised the rapid disposal of plastic bottles ("Ex und hopp").[19] Even the traditional deposit system of glass bottles was threatened in 1967 when trading companies in the beverage industry pronounced the abandonment of the system because it was cheaper to manufacture new bottles than to collect and reuse old ones.[20] The concept of the "lost package,"[21] as the contemporaries called it, was carried to the extremes.

Figure 8.1. Production of Plastics in West Germany

Year	Plastics Production (in millions of tons)
1950	0.379
1955	0.384
1960	0.982
1965	1.999
1969	3.963

Source: Deutscher Bundestag/ Drucksache VI/1519, 4 December 1970, Koblenz Federal Archive, B 106, 29370.

Figure 8.2. Gross Production Value of West Germany's Packaging Industry (1000 DM)

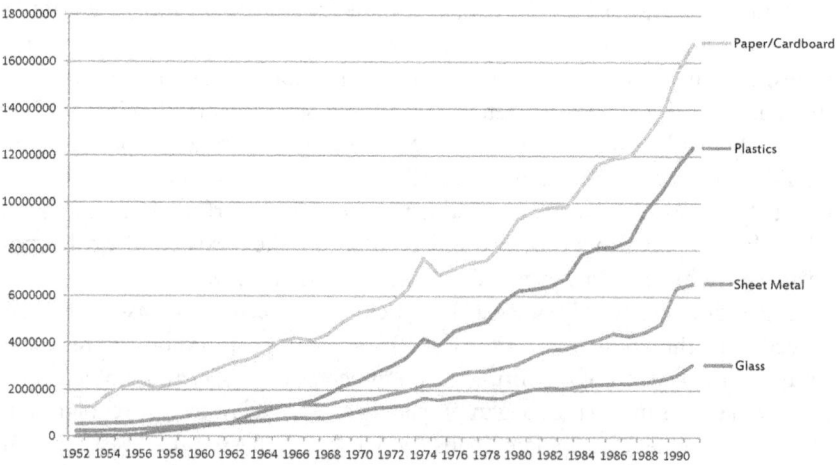

Source: Matthias Nast, *Die stummen Verkäufer. Lebensmittelverpackungen im Zeitalter der Konsumgesellschaft. Umwelthistorische Untersuchung über die Entwicklung der Warenverpackung und den Wandel der Einkaufsgewohnheiten (1950er bis 1990er Jahre)* (Berlin, 1997), 180–181.

The rising amount of plastic bottles and packaging was an urgent problem because public authorities were faced with a severe disposal problem caused by an incremental scarcity of landfill space. In highly populated areas such as the Ruhr district, authorities set up a commission in the late 1950s to solve the region's disposal problems.[22] The law protecting water reservoirs (*Wasserhaushaltsgesetz*), which came into effect in 1960, made roughly 80–90 percent of all landfills and tips illegal.[23] And even though hardly anyone in North Rhine-Westphalia dared to interpret this law literally, it was clear that something had to be done, not least because of the continually growing waste stream since the Second World War, a development that was not expected to stop.[24]

Therefore, the abandonment of the glass bottle deposit system would have caused a rise in the amount of waste of up to 30 percent from one day to the next.[25] In the 1960s, many German cities invested a great deal of money in new incinerators, facilities that often turned out to be more expensive and less reliable than initially expected. For these cities, such a measure would have meant the breakdown of their disposal systems, especially because glass did not burn very well. The constantly rising production of plastics made the disposal problems even worse. In the early 1960s, *Müll-Lawine* (waste avalanche) became a commonly used term.[26]

These incidents created a situation where it seemed reasonable to put recycling back on the agenda, at least to slow down the growth of the waste stream. When the German state began preparing the aforementioned law to regulate waste management, the officials perceived recycling as a way to reduce the amount of waste and to save energy, which, from their point of view, was squandered in plastic manufacturing. In 1973, the Ministry of the Interior launched an autonomous recycling program that later became part of the government's *Abfallwirtschaftsprogramm* in 1975.[27] For a society with a growing awareness of the threatening consequences of an ever-growing waste stream and a developing sense of environmental protection, recycling became a major issue, especially in terms of creating an alternative to tips and incinerators. The fundamental problem, that consumer society's recognized wealth and affluence had side effects that could not be avoided, forced critical protagonists to offer alternative solutions.[28] Besides composting, recycling appeared as the main alternative to traditional waste-disposal methods.

Revival of Recycling in the 1970s: When Market Solutions Failed

Even though there was a growing acceptance and an emerging public consensus that, in principle, recycling was a good thing, putting it into practice was another matter. The first private attempts, for instance the collection of old telephone books by church and welfare organizations in 1969, were nothing more than a drop in the ocean.[29] The collection of paper, glass, metals, and textiles (at this time there were no reasonable recycling facilities for plastics) had to be organized and someone had to turn it into a business. This difficulty overshadowed recycling activities in the 1970s, causing numerous failures.

The initial problem was that public cleansing departments could not provide a recycling service themselves. Since the late 1950s, local authorities had been struggling with a massive scarcity of workforce and strived therefore for a rationalization of the service, which meant new vehicles, bin types, and staff cutbacks, all for a steadily growing waste stream. To instigate separate collections of glass, paper, and so on would have required massive additional investments and no one knew where the money would come from. These investments also would have caused an additional financial burden on citizens; fees for waste collection and disposal were always a highly sensitive matter in city politics.[30]

The solution, then, was to hire private contractors. For them, the collection of waste paper and glass made sense for two reasons: Despite the increasing use of plastics, the consumption of glass and paper had been constantly growing after the Second World War.[31] Technical improvements increased the proportion of waste paper processed in paper manufacturing.[32] This strengthened the secondary market and led to a rise in prices, especially in 1973.[33] The other incentive for private contractors was that recycling seemed like an opportunity to break into the field of public services, which had been a "closed shop" until this time. The idea was that private firms should collect the materials and sell them at a price that enabled them to make a profit. Collection was voluntary. People brought their glass and paper to collection points or placed them in a bundle in front of their houses. The challenge was to make the collection profitable, and this is why recycling remained a relatively minor business in the 1970s: the secondary markets for valuable waste materials were so volatile and incalculable that it was impossible for most contractors to sustain the service.

Case Study: Recycling Waste Paper

A clear example of the problems involved in setting up a regular recycling service is the waste paper market. As already mentioned, the industrial recycling of waste paper was important in Germany; waste paper made up nearly 50 percent of paper and cardboard production. Considering the constantly rising consumption of paper in West Germany (see figure 8.3), it was obvious that a secondary market for waste paper held promise.

The increase in production and consumption of paper and cardboard was faster than the growth of the gross national product. Simultaneously, the tech-

Figure 8.3. Consumption of Paper and Cardboard in West Germany, 1950–1970

Year	Consumption of paper and cardboard (1000 tons)	Index	kg per capita	GNP in 1962 (billion DM)	Index
1950	1604	100	32.2	139.8	100
1951	1843	115	36.5	155.0	111
1952	1831	114	36	169.0	121
1953	2179	136	43.9	182.4	130
1954	2571	160	50.5	195.5	140
1955	2857	178	55.4	219.1	157
1956	3048	190	58.6	234.4	168
1957	3346	209	63.5	247.9	177
1958	3510	219	66.0	256.1	183
1959	3843	240	70.6	273.7	196
1960	4396	274	79.3	328.4	235
1961	4592	286	81.4	346.2	248
1962	4810	300	84.6	360.1	258
1963	5025	313	87.2	372.5	266
1964	5522	344	94.7	397.3	284
1965	5791	361	101.5	419.5	300
1966	5884	367	101.8	431.7	309
1967	5697	354	98.2	430.5	308
1968	6674	416	110.9	461.7	330
1969	7483	467	122.5	499.1	357
1970	7641	476	125.0	523.4	374

Source: Bundesministerium des Inneren, *Verwertung von Altpapier*, 18.

nology of paper recycling became more energy efficient and achieved better quality. Conversely, price fluctuations made this market extremely shaky (see figure 8.4).[34]

Figure 8.4. Prices for Waste Paper (Average per Year)

Year	Price index (1962 = 100)	Year	Price index (1980 = 100)
1968	109.2	1978	58.5
1969	117.3	1979	82.6
1970	138.6	1980	100
1971	99.9	1981	61.6
1972	92.6	1982	59.4
1973	106.3	1983	59.6
		1984	81.2

Source: Rationalisierungs-Kuratorium der deutschen Wirtschaft, *Verpackung und Recycling* (Frankfurt, 1975), 17–18.

These fluctuations were influenced by many factors: overall demand, market relations, transport costs, number of suppliers, general paper prices, and the appearance of new competitors in the world market. Another reason was that paper collection fell prey to its own success—an effect that would later on also harm the so-called dual system (*Grüner Punkt*) of recycling, which was introduced in 1990. The simple logic behind this was that the expanding paper collection led to a rise in paper supply, thereby causing a drop in prices. In other words, occasionally too much paper was collected to achieve a reasonable price. Such inconsistently profitable collections of waste paper were not good preconditions for the establishment of a regular recycling service.

A paradox situation arose after the mid-1970s. There was a broad acceptance of recycling, not least in light of the continually growing waste stream (a relative stagnation of the amount of household waste did not occur until the 1980s).[35] However, there were numerous obstacles to the installation of a regular service, especially price fluctuations in the secondary markets, as well as the lack of a well-organized system—which for the time being consisted of merely a few collection points and special bins for paper and glass. Practitioners were aware that "bring-systems" whereby people had to bring their waste to a specific place, suffered from a lack of acceptance and convenience, and also caused "social problems" when containers were set on fire or when they became the scene of small social riots.[36] Collections were rarely frequent and covered only certain areas. In the late 1970s, it became clear that market solutions were insufficient to cope with the recycling issue.

State-Fostered Recycling since the Late 1970s

In the late 1970s, there was not only (in principle) a broad acceptance of recycling by the public, but also growing pressure from citizens' initiatives to extend recycling activities and prevent the use of potentially harmful materials for packaging and other purposes.[37] In 1977, for instance, some of these campaigners met in front of the Ministry of the Interior in Bonn, disposing wagons of aluminum cans (although some of these were actually made of tin plate).[38] Cities were forced to publish booklets providing evidence of their activities concerning improved protection of the environment.[39]

The way to solve the failing market solutions was to guarantee prices. Many German cities hired contractors in the late 1970s and guaranteed to pay them the difference when the price fell under a certain level. On the other hand, if the price rose over this level, the contractors would pocket the difference as normal. In this way, cities created an incentive to install (and could be held liable to provide) a regular service.[40] Private firms were well-prepared for this task. Waste management companies in the private sector experienced such a steep rise in the 1970s that by the end of that decade, they were collecting and disposing of the waste of more than half of the people in West Germany. Furthermore, they had built up their own infrastructure, including landfills, incinerators, and recycling facilities.[41]

These were the preconditions for recycling becoming a regular service in the 1980s. At this time, all the large cities in Germany installed collection points for paper or started a regular collection of waste paper.[42]

Figure 8.5. Return Rate for Scrap Glass

Year	Returned Scrap Glass
1974	150,000 tons
1975	200,000 tons
1976	260,000 tons
1977	310,000 tons

Source: "Memorandum des Bundesverbandes der deutschen Industrie zu den Entwicklungslinien der deutschen Abfallwirtschaftspolitik vor dem Hintergrund des deutschen Abfallwirtschaftsprogramms" (July 1978), Koblenz Federal Archive, B 106, 69723.

Local authorities had to provide evidence of their recycling activities to a concerned public. Bigger cities could no longer lag behind. The renewed waste management law passed in 1986, as well as the packaging regulation (*Verpackungsverordnung*) thereafter, prioritized recycling.[43] Additionally, this legislation strongly recommended contracting private firms, demonstrating once again the close link between the expansion of recycling and the increasing importance of private firms in the waste management industry.

Meanwhile, consumer behavior was also changing. For a long time, introducing new technologies in waste collection had been such a huge obstacle that practitioners assumed people would not accept them. They had reason to think so: several attempts to rationalize waste collection had failed because of people's "misbehavior."[44] Recycling, though, was a different matter. People proved to be willing to sort their waste and to carry their glass bottles to collection points. This certainly stemmed from an ecological awareness that had developed over the past twenty years. Recycling was significant in that it offered people an opportunity to engage actively in environmental protection, while air and water pollution or forest dieback seemed to be beyond the power of individuals. Local authorities' efforts to educate people to behave responsibly also played a role.[45]

The earlier-mentioned state-funded dual system introduced in 1990 marked the beginning of a new era in German recycling history. It was, however, not mainly dedicated to mitigating the scarcity of raw materials (as plastics are not scarce) or decreasing energy costs, but rather was a reaction to a severe "waste crisis" since the mid-1980s. At this point in time, the disposal infrastructure, created in the aftermath of the waste law from 1972, was about to reach its limits. The dual system was intended to reduce the amount of plastics in the waste stream, the third main fraction of materials after glass and paper.[46] However, it demonstrates the extent to which the introduction of the German recycling regime came as a result of the very practical problem of scarce disposal space. It was therefore no actual "alternative" to careless consumption, but more or less a measure to deal with its consequences.

Furthermore, the dual system faced considerable problems because those in charge had massively underestimated the willingness of people to sort their waste. For this and other reasons, companies responsible for organizing recycling struggled to cope and some even went bankrupt by the early 1990s. But this situation did not endanger the German recycling regime as a whole, which had begun to develop in the 1980s.[47] It required close cooperation between public authorities and private contractors, efficient markets for secondary raw materials, and a general public willing to sort waste and respect recycling efforts. This system was economically guaranteed by the German state, which, far from hindering, in fact enabled and fostered a market solution.

Conclusion

This chapter aims to outline why and how Germany's recycling system evolved. My argument is that the emergence of modern recycling in the late 1960s was the result of a juxtaposition of a steeply growing waste stream, a lack of disposal facilities, an energy crisis on the horizon, emerging markets for secondary raw materials, and an evolving awareness of the necessity of environmental protection. Despite this situation, it took more than fifteen years before a functioning recycling system was put into practice, albeit one that would serve as model for recycling activities in many other industrialized countries.

Obviously, there is no monocausal explanation for the emergence of recycling in West Germany. The questions we still need to ask are whether some factors were more important than others and to what extent the economic dimension of recycling was significant. This chapter argues that the main drivers of "old-fashioned" recycling were economic ones; when city planning, self-service stores, etc., reduced the opportunities for "convenient" recycling, it stopped, mainly because it was no longer worthwhile. As long as there were small and regional markets for secondary raw materials, the expanding mass markets made recycling unprofitable. The declining production costs for manufacturing glass and paper also played a role. On the other hand, the emergence of "new" recycling was also fueled by economic incentives. Without expanding the mass markets for secondary raw materials, it would have been much more complicated to instigate a regular recycling service, even though pure market solutions failed in Germany in the 1970s.

So how, after all, can we distinguish "traditional" from "modern" recycling? The first difference in economic terms is the extension of markets. In earlier times, paper and metals were sold on small and regional markets, mostly to local traders. Modern recycling depends on much bigger markets, and today there are globalized markets for secondary raw materials. The second difference is that recycling is no longer driven by the absolute scarcity of raw materials, but on the basis of price differences guaranteed by the state or municipalities. The demand for paper and glass, for instance, could principally be satisfied by other means or could be substituted by materials like plastics. The third difference is the crucial role of the state. Mainly to diminish the amount of waste and for environmental protection, cities foster the markets by guaranteeing prices and making the business of recycling sustainable, without organizing recycling activities independently. On the contrary, recycling was one of the first areas to be privatized in the German economy after the Second World War.

This brings another issue to the fore, concerning environmental history as well as economics. For environmentalists, industry is often generally seen as bad insofar as it represents the main polluter. Its protagonists oppose environmental protection because it reduces their profits. Environmental economics, on the other hand, normally operate within the contradiction of state vs. market, whereby state regulation generally proves to be inefficient. Environmental policy goals should be to create market solutions that achieve more efficient outcomes. The case of recycling expounds the dilemma of this contradiction. The state did not hinder market solutions, but quite the opposite: By demonstrating that not everything that makes a profit is desirable (e.g., substituting glass with plastics), the state was defining the institutional framework for recycling. It "created" markets by regulation and guaranteeing prices.

When recycling in West Germany emerged as a "role model" for countries like the United States and Britain after the late 1960s, it was probably not only thanks to the fact that this country brought goods back into a product "lifecycle." The West German case shows that regulation does not necessarily kill market solutions, but can foster and create them. This was certainly something that could be learned from the German example and has had a deep impact on environmental protection policies up to this day.[48]

Roman Köster is senior lecturer at the Department for Economic and Social History at the Bundeswehr University in Munich, Germany. After finishing his studies at the University of Bochum in history and German literature, he received his PhD from the University of Frankfurt in 2008. He has published books on German business history in the twentieth century and the history of political economy during the Weimar Republic. Furthermore, he is coauthor with Raymond G. Stokes and Stephen Sambrook of *The Business of Waste*, a volume on the history of waste management in Britain and West Germany after the Second World War. His research interests are economic history, environmental history, and the history of economic thought.

Notes

1. See, for instance, Martin Melosi, *Garbage in the Cities: Refuse, Reform, and the Environment*, revised edition (Pittsburgh, PA, 2005), 237. Louis Blumberg et al., *War on Waste: Can America Win Its Battle with Garbage?* (Washington, DC, 1989); Finn Arne Jørgensen, *Making a Green Machine: The Infrastructure of Beverage Container Recycling* (Baltimore, MD, 2011).
2. Ruth Oldenziel and Heike Weber, "Introduction: Reconsidering Recycling," *Contemporary European History* 22, no 3 (2013): 347–370; Raymond G. Stokes, Roman Köster, and Stephen Sambrook, *The Business of Waste: Britain and Germany, 1945 to the Present* (Cambridge, 2013). See also the special issue on recycling, *Technikgeschichte* 71, no. 1 (2014), ed. by Heike Weber.
3. Peter Münch, *Stadthygiene im 19. und 20. Jahrhundert. Die Wasserversorgung, Abwasser- und Abfallbeseitigung unter besonderer Berücksichtigung Münchens* (Göttingen, 1993); Sonja Windmüller, *Die Kehrseite der Dinge. Müll, Abfall und Wegwerfen als kulturwissenschaftliches Phänomen* (Münster, 2004).
4. Münch, *Stadthygiene*.
5. Josef H. Bausch et al., *Es herrscht Reinlichkeit und Ordnung hier auf den Straßen. Aus 400 Jahren Geschichte der Stadtreinigung und Abfallentsorgung in Dortmund* (Dortmund, 2005), 54–55.
6. See Friedrich Huchting, "Abfallwirtschaft im Dritten Reich," *Technikgeschichte* 48, no. 3 (1981): 252–273.
7. M. A. Jinhee Park, *Von der Müllkippe zur Abfallwirtschaft. Die Entwicklung der Hausmüllentsorgung in Berlin (West) von 1945 bis 1990* (Berlin, 2004), 24–27; Hildegard Frilling and Olaf Mischer, *Pütt un Pann´n. Geschichte der Hamburger Hausmüllbeseitigung* (Hamburg, 1994), 135–141; Malte Zierenberg, *Stadt der Schieber. Der Berliner Schwarzmarkt 1939–1950* (Göttingen, 2008).
8. "Altpapier landet tonnenweise auf dem Müll," *Stuttgarter Zeitung*, 9 January 1974, Mannheim City Archive, Bauverwaltungsamt, Zugang 52/1979, Nr. 950.
9. Abfallwirtschaftbetriebe Köln, ed., *111 Jahre Abfallwirtschaft in Köln* (Cologne, 2001), 75–76.
10. Hans Baumann, "Städtereinigung. Hygienische Aufgaben des Alltags," *Der Städtetag* (May 1960), 257–258.
11. For the German case, see Michael Wildt, *Am Beginn der Konsumgesellschaft. Mangelerfahrung, Lebenshaltung, Wohlstandshoffnung in Westdeutschland in den fünfziger Jahren* (Hamburg, 1994).
12. These letters, however, were in most cases not written by housewives, but by their husbands. But it is hard to see how this could have had any impact on the lack of recycling issues.
13. See: Staffan Linder, *Harried Leisure Class* (New York, 1970).
14. Christian Pfister, "The '1950s Syndrome' and the Transition from a Slow-Going to a Rapid Loss of Global Sustainability," in *The Turning Points in Environmental History*, ed. Frank Uekötter (Pittsburgh, PA, 2010), 90–118. It also fits with the concept of the "Great Acceleration"; see this volume's introduction.
15. Bundesministerium des Inneren, ed., *Verwertung von Altpapier. Untersuchung über die*

Möglichkeit der Verwertung von Altpapier. Gegenwärtiger Stand und zukünftige Entwicklung. Bericht des Battelle-Instituts Frankfurt am Main (Berlin, 1973), 14.
16. Ibid.
17. Anonymous letter to City of Mannheim (22 July 1971), Mannheim City Archive, Bauverwaltungsamt, Zugang 52/1979, Nr. 1463; letter from Heinz Berchtold to City of Mannheim (3 September 1971), Mannheim City Archive, Bauverwaltungsamt, Zugang 52/1979, Nr. 1463.
18. Bundesministerium des Inneren. Projektgruppe Abfallbeseitigung. Broschüre: Brennpunkt Müllproblem (Bamberg, 1968), Koblenz Federal Archive, B 106, 29370; letter from Tiefbauamt to Dezernat VII (29 June 1967), Mannheim City Archive, Bauverwaltungsamt, Zugang 52/1979, Nr. 1463.
19. See Andrea Westermann, *Plastik und politische Kultur in Westdeutschland* (Zurich, 2007).
20. "Wohin mit dem Wohlstandsmüll? Einwegflaschen kosten Stadtreinigung zusätzlich mehrere Millionen," *Mannheimer Morgen,* 12 April 1969, Mannheim City Archive, Bauverwaltungsamt, Zugang 52/1979, Nr. 940.
21. For instance, letter from Bürgermeisteramt Stuttgart to Städteverband Baden-Württemberg (18 May 1967), Mannheim City Archive, Zugang 52/1979, Nr. 1463. "Ergebnis der Ermittlungen über Anfall von PVC-Abfällen und Einwegflaschen sowie deren Auswirkung auf die Abfallbeseitigung," July 1969, Koblenz Federal Archive, B 106, 29370.
22. Letter from Bundesverband der Deutschen Industrie, Landesvertretung Nordrhein-Westfalen to Minister für Landesplanung, Wohnungsbau und öffentliche Arbeiten des Landes NRW Joseph Franken, 5 May 1965, Düsseldorf State Archive, NW 354, 586.
23. Ibid.
24. Stellungnahme Deutscher Rat für Landespflege: Zum Problem der Behandlung von Abfällen, October 14/21, 1970l, Mannheim City Archive, Bauverwaltungsamt, Zugang 52/1979, Nr. 944.
25. Letter from Tiefbauamt to Dezernat VII (29 June 1967), Mannheim City Archive, Bauverwaltungsamt, Zugang 52/1979, Nr. 1463.
26. For example, "Die Konjunktur hat auch eine Kehrseite: Ersticken wir im Wohlstandsmüll?," *Rhein-Neckar-Zeitung,* 1 September 1960; "Die Müll-Lawine überrollt alles. 50.000 Ablagerungsplätze in der Bundesrepublik sind überschwemmt," *Süddeutsche Zeitung,* 9 December 1970, Mannheim City Archive, Bauverwaltungsamt, Zugang 52/1979, Nr. 940.
27. Statement Referat UB I 6. "Ergebnisvermerk Recyclingprogramm der Bundesregierung. Bezug: Sitzung am 11.7.1973" (16 July 1973), Koblenz Federal Archive, B 106, 58783.
28. Ecosystem. Gesellschaft für Umweltsysteme mbH, "Vorstudie Abfallwirtschaftsprogramm der Bundesregierung im Auftrag des Bundesministerium des Innern," Bonn, December 1973, Koblenz Federal Archive, B 106, 58783.
29. "Stellungnahme Deutscher Rat für Landespflege: Zum Problem der Behandlung von Abfällen" (21 October 1970), Mannheim City Archive, Bauverwaltungsamt, Zugang 52/1979, Nr. 944.
30. For instance, "Auszug aus der Niederschrift über die Sitzung des Verwaltungs- und Finanzausschusses," 9 July 1963, Mannheim City Archive, Hauptregistratur, Zugang 42/1975, Nr. 2401.

31. Nast, *Die stummen Verkäufer,* 180f., 342f.
32. Bundesministerium des Inneren, ed., *Verwertung von Altpapier* (Bonn, 1975), 18.
33. Letter from City of Stuttgart to Gottfried Hösel, Bundesinnenministerium des Inneren (13 May 1974), Koblenz Federal Archive, B 106, 65268.
34. Wolfgang Schneider, *Sekundärrohstoff Altpapier. Markt und Marktentwicklung in der Bundesrepublik Deutschland* (Dortmund, 1988), 157.
35. Statistisches Bundesamt, ed., *Statistisches Jahrbuch der Bundesrepublik Deutschland 1982* (Wiesbaden, 1983), 561; Statistisches Bundesamt, ed., *Statistisches Jahrbuch der Bundesrepublik Deutschland 1993* (Wiesbaden, 1994), 730.
36. "Große Pleite mit Sperrmüllbehältern," *Rhein-Neckar-Zeitung* (May 12, 1966), Mannheim City Archive, Hauptregistratur, Zugang 42/1975, Lfd.-Nr. 1084.
37. See Hessischer Minister für Arbeit, Umwelt und Soziales, ed., *Hessen geht neue Wege in der Abfallwirtschaft* (Wiesbaden, 1985).
38. Letters in Koblenz Federal Archive, B 106, 69733.
39. "Umweltschutzmaßnahmen des Stadtreinigungsamtes der Stadt Dortmund," Dortmund, 1985, unpublished report, Dortmund City Archive.
40. Karl Pulver, "Von der Abfuhranstalt zum Eigenbetrieb. 125 Jahre Stadthygiene in Mannheim" (unpublished manuscript, Mannheim City Archive, 2005), 90–92.
41. Meeting minutes 28/6/1978 Umweltbundesamt (16 August 1978): Zusammenarbeit zwischen VPS und dem UBA, Koblenz Federal Archive B 106, 69732.
42. See reports in Mannheim City Archive, Bauverwaltungsamt, Zugang 52/1979, Nr. 1463; Park, *Müllkippe,* 111–116.
43. "Abfallgesetz. Bericht der Bundesregierung über den Vollzug des Abfallgesetzes vom 27. Aug. 1986" (Bonn, 1986).
44. "Große Pleite mit Sperrmüllbehältern," *Rhein-Neckar-Zeitung* (12 May 1966), *Mannheim City Archive, Hauptregistratur, Zugang 42/1975, Nr.1084.* For another example, see letter from residents of Maudacher Straße 4,6,8,10 to Städtisches Tiefbauamt, Direktor Borelly (15 March 1958), Mannheim City Archive, Hauptregistratur, Zugang 40/1972, Nr. 291. Ecosystem, Gesellschaft für Umweltsysteme mbH, Vorstudie Abfallwirtschaftsprogramm der Bundesregierung im Auftrag des Bundesministerium des Innern, Bonn, December 1973, Koblenz Federal Archive, B 106, 58783.
45. Arbeitsgemeinschaft Fichtner/ifeu. Stuttgart/Heidelberg im Dezember 1989, Abfallwirtschaftliches Gesamtkonzept für die Stadt Dortmund, Dortmund City Archive. For a critical point of view, see Windmüller, *Kehrseite,* 43–44.
46. Cf. Jørgensen, *Making a Green Machine.* Vending machines were also installed in Germany during the 1980s. They were not very common, though, and attempts to install deposit systems also for plastic bottles simply did not work.
47. Internationales Symposium ISWA/VKS, *Abfall im Wandel der Zeit* (Munich, 1987), 22.
48. Mathew Gandy, *Recycling and the Politics of Urban Waste* (London, 1994).

 CHAPTER 9

Ecological Modernization of Waste-Dependent Development?
Hungary's 2010 Red Mud Disaster

Zsuzsa Gille

On 4 October 2010, 700,000 cubic meters (24.7 million cubic feet) of toxic sludge escaped from pond no. 10, a reservoir of an alumina factory owned by the Magyar Alumínium Termelő és Kereskedelmi Zrt. (MAL Ltd.) in western Hungary. The stream, which flooded three villages, was a 25 kilometer long, 1–2 kilometer wide, and occasionally over 2 meter high cascade of mud. Ten people died from burns or drowning, and hundreds were treated in hospitals. Over two hundred houses subsequently had to be demolished. All life in the nearest river, the Marcal, a tributary of the Danube, was extinguished—fish, birds, insects, and plants. According to government officials, this was Hungary's worst ecological disaster ever.

Cases like this—tragic narratives of industrial hubris—sit uneasily with the celebratory news about industry tending toward zero waste and the strides modern societies have made in reusing and recycling. This is for two reasons. First, much of the mainstream media and industry journals focus only on relatively nontoxic by-products, and the successes in reusing and recycling them. Second, studies of recycling—whether in journalism or in scholarly writings—tend to concentrate on household waste, ignoring the fact that what we commonly refer to as consumer waste comprises only 3–10 percent of all waste generated.[1] That is, there is a huge gap between the amount of attention we dedicate to individual waste practices and their significance in the overall waste streams of modern societies.

This disaster, however, highlights another conundrum. While it is true that the concept of recycling originated in the chemical industry, as the introduction to this volume reminds us, the appropriation of this concept by environmentalists, and its revalorization as a practice of sustainability, has not meant that industry representatives have given up on its use. In fact, this case study shows that practices that, from the perspective of production technologies, will appear as closing material cycles—and thus seemingly consistent with reuse and recycling—in practice are dependent on open material cycles and end-of-pipe technologies. This doesn't simply suggest that recycling, just like the concept of waste, is in the eye of the beholder, as social constructionists would insist; but that the very materiality of recycling is more complex than the word *cycle* may lead us to believe.

In what follows, I will review the historical context of this disaster, focusing first on the company and its technology, including the causes of the spill; and second, on changes in related environmental policies, from state socialism through EU accession. In conclusion, I will evaluate which theory of the waste-economy relationship best describes this case: ecological modernization or waste-dependent development.

Red Mud

Red mud is the by-product of alumina (aluminum oxide) production—that is, of the first step in producing aluminum. In the Bayer process, bauxite is treated with a caustic solution (sodium hydroxide, or NaOH) to remove impurities. This technology was developed in the late nineteenth century and, amazingly, it is still the most commonly used process. The resulting waste has a low solid material content and is highly alkaline. The scientific scholarship places the usual pH level at 12, but in Hungary the spilled slurry had a pH of 13–14, which is a significant difference, due to the logarithmic nature of the pH scale.[2] Red mud usually contains iron oxide (rust), aluminum oxide, calcium oxide, titanium oxide, sodium oxide, and silicon oxide. While iron oxide is always in the highest concentration—lending red mud its color—every plant's by-product will have a somewhat different composition.

Depending on the quality or purity of the ore used, the amount of red mud relative to the amount of alumina produced can vary between 0.3 and 2.5 tons per ton of alumina.[3] The average ratio is one to two tons of red mud per ton of alumina produced, rendering the Bayer technology a rather wasteful production method. Among worldwide industrial processing wastes (not counting tailings from mining), red mud ranks second, constituting 23 percent of the total, following only steel manufacturing at 70 percent (Cooling 2007). According to current estimates, by 2015 there will be 4 billion tons of bauxite residue (red mud) globally.[4]

Two main disposal methods are in use currently. Where there is coastal access, it is dumped into the sea, a practice followed in at least France, the United Kingdom, Germany, Greece, and Japan.[5] More land-locked facilities instead have created lagoons or holding ponds for the sludge, which can then be dewatered and treated to a varying extent. Hungary, with no access to the sea, has been using this method for many decades. Until relatively recently, there were three areas near various alumina facilities that hosted such lagoons: Almásfüzitő, near the Slovakian border; Mosonmagyaróvár, near the Austrian border, a site that has now been closed and reclaimed; and Ajka, in the northwest of the country, 50 kilometers north of Lake Balaton, the region that was the source of the 2010 disaster.

Wet deposition—that is, the lagoon method—is essentially the traditional way of dealing with tailings from mines, and it is the cheapest way to dispose of red mud. One shortcoming of this process is the high probability of seepage into the soil and groundwater. Another is the corrosion of the lagoon walls due to the low viscosity of red mud and the continuous movement of the sludge in the impoundment (due to rain and wind), which can lead to the fracture of the walls and consequent flooding. What renders these two scenarios of red mud escape particularly dangerous is the high causticity (alkalinity), rather than the toxicity of the material.

Recognizing the risks, starting around 1980, most Western alumina factories shifted to disposal methods that reduce both the wetness and the alkalinity of red mud. Most common has been the so-called dry stacking method, in which underground drainage and the sloped spreading of the red mud allows it to dry to 70 percent solid content, after which another load can be piled on top. Increasing the solid content to an even higher level does not seem to be desirable, however, because winds can stir up red mud dust, which presents a serious health hazard. The advantage of dry stacking is that the removal of the caustic liquid from the by-product also decreases its alkalinity. Other methods of neutralization include using seawater, which brings the pH level down to a range of 8.0–9.5. The problem with this method, however, is that it introduces salt and potentially other impurities into the leachate, which can endanger agriculture and drinking water sources. A more recent method that avoids salination and can reduce global greenhouse gas emissions is carbonation. Here the sludge is suffused with carbon dioxide gas, bringing alkalinity down to a pH range of 9–10. Both of these newer methods are considered expensive, though such expenditures are likely to be offset by savings on future remediation.

The Hungarian Aluminum Industry and the History of Red Mud

Ajka is home to the largest alumina-producing facility in Hungary, established in 1941. The Hungarian aluminum industry, however, only developed significantly after the communist takeover in 1948. The new government, in view of Hungary's lack of raw material resources necessary for heavy industry and energy generation, jumped at the chance to utilize the significant bauxite sources concentrated in the northwestern quadrant of the country. As an industry of utmost military and economic significance, its enterprises (the bauxite mines, alumina factories, smelters, and mills) were not only nationalized but, in addition, were subsumed by a Hungarian-Soviet "joint venture" called Maszobal in 1950.[6] Given Hungary's high share in the total bauxite production of the socialist bloc, peaking at one-third, such an economic "cooperation" was to be expected. The Soviet need to keep tabs on such an important sector didn't change, but how the Soviet Union did so changed over time. While the Hungarian government bought out the Soviet partner in 1954, the most energy-intensive part of the production process was farmed out to the Soviet Union, where energy needed for smelting was more abundant and cheaper. According to this deal, signed in 1962, the difference in price between the alumina Hungary "exported" and the aluminum "imported" back was paid either in Hungarian industrial products or in extra alumina.[7] The two countries only agreed to end this barter in 1988, intending to transform it starting in 1990 into a more conventional business contract, with prices determined by the world market and the balance settled in cash. The collapse of state socialism in Hungary in 1989, however, radically changed the landscape of economic cooperation.

Privatization and Ownership of the Red Mud Lagoons

In 1997, the Ajka Alumina Factory (*Ajkai Timföldgyár*) became the property of the Magyar Alumínium Termelő és Kereskedelmi Zrt. (MAL Ltd.), a private corporation that eventually vertically integrated almost all Hungarian bauxite and alumina processing, and later even bought other Eastern European plants. Privatization didn't take place overnight, it was rather a multi-staged and complicated process that started in 1995, and was implemented unevenly in the various segments of the industry. While initially the state resisted privatization, due to the importance of the industry for Hungary's trade balance, the first half of the 1990s saw a severe drop in aluminum prices, which suddenly tipped the balance in favor of selling. However, the state did not give up all interest in the future of the industry: the privatization agency in charge of the sale gave preference to Hungarian bidders over foreign ones, with the argument that foreign owners would only want to close these factories and flood the market with their products manufactured elsewhere, a practice that by then had become common all over the postsocialist world. In fact, one of the conditions of the privatization was that the company would continue production and not lay off people for the foreseeable future. The new owner, Árpád Bakonyi, was the former board chairman, appointed by the state just before the privatization commenced—a process the scholarship calls "managed privatization."[8]

While the privatization contract demanded that MAL carry out environmental amelioration and implement ecological innovation in its technology, due to its running deficit in most of the early 2000s, MAL didn't comply. In the first postsocialist Hungarian privatizations up to 1992, the new owner typically received a guarantee that the state would clean up whatever environmental damage was later discovered. In some cases, this cost exceeded the price of the firm, thus reducing the state's privatization income to zero. In addition, the amount and time limit of this guarantee were not subject to expert assessment but simply based on a bargain between seller and buyer. While a 1992 law ended this situation, it did not apply to so-called partial privatizations. In these, the most profitable and cleanest plants of state enterprises were privatized, while those with deficits and environmental problems were left in the hands of the state, thus leaving the cleanup costs to those units that could least afford to pay them.

The complex process of privatization in the Hungarian alumina industry created a situation of partial privatization, as a result of which the red mud ponds that had been decommissioned remained in state ownership, while the ones still in use, such as those around Ajka, were transferred to the new owner. The contract estimated the environmental remediation costs of these transferred ponds at 3.3 billion forints, which was used to justify the ridiculously low sale price of 10 million forints for assets estimated to be worth 1591 billion forints.[9] Intentionally undervaluing assets to be privatized in return for kickbacks was another common practice in postsocialist denationalization, a likely cause of the low sale price in this case as well. The contract obliged the new owner to send annual reports to the privatization agency on the progress made in environmental remediation, and threatened financial sanctions should the new owner fail to invest the prescribed sum in environmental projects. This penalty, however, was again so low (10 percent of the value of the damage caused) that it could not help but make it rational for MAL not to fulfill the terms of the contract. So let us see what of the stipulated environmental investments MAL actually did complete.

There are ten holding ponds adjacent to the Ajka factory, and they were all established under state socialism, when the factory was state owned. The lagoons were meant to store red mud and grey sludge. (The latter is the tailing resulting from coal mining, and it was stored here because Ajka's power plant burns locally mined coal.) At the time of the disaster, there was an estimated volume of 50 million cubic meters of grey and 30 million cubic meters of red sludge stored in the ponds, and grey sludge had been used for building the impoundment walls of the more recent lagoons. Pond no. 10 was the tallest, with walls as high as 20–25 meters. As mentioned above, the key environmental hazard of lagoons such as these is their seepage, which indeed had been constant, according to nearby villagers I interviewed. Another potential hazard is that sludge with high water content (and the red mud around Ajka had only a 50–60 percent solid content) moves about and thus weakens the walls of the impoundment over time.

One environmental installation that had been prescribed by the state and that MAL did complete was an underground buffer wall that was supposed to prevent seepage into the surrounding soil. This construction began in the early 1980s, and after its completion in 2001, it surrounded ponds nos. 6–9, with a length of 7.4 kilometers. According to MAL's representatives, however, this barrier allowed water to pool at the base of the lagoon and over time contributed to weakening the dams' retaining walls. Of course, the fact that this installation was prescribed by the state allowed MAL to blame the state for the disaster. According to other experts, a more significant cause of the walls' fracture was that they had been raised multiple times, thus increasing the pressure they had to withstand. In order to understand the circumstances involving the storage of red mud, and specifically the height of the walls of pond no. 10, I have to introduce both the national and local regulations.

Regulation on Different Scales

In contrast to the mainstream belief that socialist countries had no substantial environmental policies, Hungary introduced its first hazardous waste policy in 1981, well before the collapse of the Soviet bloc. This classified waste into three categories of increasing dangerousness, and red mud was considered to be in the second category. Red mud comprised and still comprises a rather high ratio of all hazardous waste—about 25 percent in 2000—which renders Hungary's waste composition unique among all industrial countries.[10]

Until 1979, the red mud around Ajka was neutralized by dewatering it, which made it a fairly solid material that could be moved by shovel.[11] The amount of dry red mud transported by trucks to the disposal sites became prohibitively large, and thus very expensive, at a time when energy prices were already at a record high; consequently, the dewatering process was stopped in favor of piping the more liquid wastes directly into the newer ponds. This, in some sense, brought or promised to bring an improvement in air quality, much spoiled by the dust blowing off the drying red mud sites. A Hungarian environmental sociologist, Viktória Szirmai, documented these problems in a report citing both her informants and contemporary newspapers writing about a veritable red cloud hovering over the land, caused by the dust from the drying red mud.[12] According to her sources, visibility during windy weather was reduced to 4–5 meters.[13] But the newer ponds exacerbated another problem: leakage that, as tests would show, rendered drinking water from nearby wells unfit for human consumption, and caused plants and trees to wither.

Szirmai also analyzed the debates between an ad hoc committee appointed by the municipal government (*városi tanács*) of Ajka and the Ajka alumina company, in which the height of the impoundment walls was the most contested issue. According to the environmental authorities, the company had a permit for a height of 16 meters, but in its plans for increasing the capacity of the holding ponds, it was referring to wall heights of 25–40 meters. Given that, according to the committee, nowhere in Europe did walls exceed 9 meters, one can understand the local concerns, among which the danger of dam failure loomed large. The local political committee of the party and the alumina factory, however, shared an interest in increasing capacity by elevating the pond walls rather than by establishing new lagoons, because the latter course would have taken huge areas of land from agricultural cultivation. So eventually, her report concludes, the municipal government stopped representing the citizenry's environmental interests and issued the permits for elevating the walls.

At the same time, the older, already full ponds were left without plant covering.[14] In addition, seepage continued through the decade, both from the newer ponds and even from the formerly dewatered ones, for example after heavy rains. Nevertheless, the fact that red mud was classified as hazardous waste allowed for stricter monitoring that, even if not religiously carried out under state socialism, provided the alumina producer with certain incentives to comply. In fact, I would argue, this was the kind of oversight that led to the formation of the ad hoc committee in the first place, and that allowed the burgeoning of new literature on hazardous wastes in Hungary that, in turn, supplied the committee with the information about other countries' practices, including the usual dam heights in Europe.

According to MAL, however, the socialist state had committed another mistake by allowing pond no. 10 to be built in that particular location. Two different kinds of soil substrates meet at the walls of this particular reservoir, causing differential reaction in the foundation to pressure and precipitation, and thus causing cracks. For years, university professors in engineering have been teaching the case of these reservoirs to their students as a negative example of siting a waste facility, not only because of the soil conditions but also because of its proximity to residential areas, to rivers that flow into the Danube and to drinking water sources.[15] However, the Parliamentary Committee investigating the disaster rejected this factor as the main cause of the dam failure.[16]

By the time Hungary was getting ready to enter the European Union in the early 2000s, alumina producers in the West European countries had mostly shifted to more advanced disposal methods that, as mentioned, resulted in a drier and less alkaline by-product. Consequently, the European Union's Waste Code does not explicitly consider red mud a hazardous waste, although other stipulations in its environmental policy lead one to conclude that highly caustic materials, such as the red mud stored near Ajka, should be considered hazardous. In preparation for adopting the Environmental Acquis, Hungary adopted the EU's waste classification, although the Hungarian government could have retained the older classification without violating EU policy.[17] Another international regulatory entity, the Basel Convention, explicitly lists red mud as hazardous, and Hungary has been a signatory of that treaty since 1989.

So why did the Hungarian environmental authorities fail to retain the hazardous waste classification for red mud? One response lies at the empirical level of regulatory changes, and many analysts of the case have focused on this. It is another complicated story, in which the timing of regulatory changes, such as the harmonization of Hungarian and EU environmental laws, created a kind of a loophole or window of opportunity that gave the task of classification to the producer. In effect, MAL ended up declaring to the Hungarian state that its red mud was not hazardous waste. This was not simply self-regulation; it was self-diagnosis or self-interested interpretation of the law.[18]

Further complicating the issue of classification is that, according to MAL, red mud is not a waste product but an intermediary material in the aluminum production process. As a result, company representatives still do not consider the ponds at Ajka waste disposal sites, but rather production sites, in which the filtering of red mud and the recuperation of the caustic NaOH solution take place; that is to say, materials are not simply moved and left there, but materials are also taken out. In its response to the Parliamentary Committee's report, MAL insists that a red mud lagoon does not become a disposal site until after it is closed. Neither is red mud toxic, according to MAL. The company and many of the experts it relies on still argue that red mud is not toxic, it is simply extremely alkaline. In Hungarian, the word *toxic* is usually translated as *mérgező*, which means poisonous. Clearly, MAL argues, red mud is not going to poison one on contact, but one should avoid skin contact with it, for example: "one should not bathe or swim in it," as the CEO said on the day of the catastrophe.[19] Whether or not we include extreme alkalinity in our definition of toxic, it cannot be doubted that red mud is hazardous.

In essence, the response to the question of why red mud was not classified as hazardous waste focuses on regulations, blaming the incompetence of Hungarian environmental authorities for (allegedly) misinterpreting international policies and laws. For some, this incompetence argument is complemented, and for others replaced, by the corruption argument—that is, the fusing of economic and regulatory power. It is true that MAL's owners had connections to the top ranks of the Hungarian Socialist Party, which has been in and out of government for the last twenty years; and, as mentioned, the managed privatizations within the alumina industry favored former ministry officials, such as Bakonyi, as well as the circle around Ferenc Gyurcsány, who served as Hungary's prime minister between 2004 and 2009. Though this first response is not incorrect from a factual perspective, in my view it shortcuts the analysis. The other answer to the hows and whys of classifying red mud lies at a deeper structural level, and I prefer to focus on this.

Structural Connections: State Capacity and For-Profit Uses of Red Mud

Key for such a perspective is analyzing privatization not simply based on the owners' identity and political connections, but rather as a deep structural transformation of postsocialist societies. Privatization from this perspective was: (a) a mechanism to redistribute wealth, (b) a process of capital accumulation, since redistribution in reality led to a concentration of assets, (c) the weakening of the state's economic power, and (d) the ascendance of the profit motive to the supreme rationale of all economic activity. These factors were combined in the red mud case. The transfer of the alumina industry into private hands weakened the government—and not only in the sense I described above, namely by shortchanging the state. Low state income from privatizations also had the cumulative effect of siphoning off resources from, among other things, regulatory enforcement. As I detailed in my book *From the Cult of Waste to the Trash Heap of History*, not only was there an overall tendency to deregulate after 1989, but the staff and monetary resources available to environmental authorities were also severely depleted.[20] Both circumstances are known from the scholarship to create favorable conditions both for regulatory incompetence and for corruption.

The simultaneous concentration of capital in turn increased the leverage of MAL's new owners, several of whom are counted among Hungary's twenty richest citizens. As we know from the literature on Western capitalist countries, such economic powerhouses can seriously impede the state's ability and willingness to monitor their activities even in the absence of explicit bribery or other forms of corruption. It is usually sufficient for such interests to cite their contributions to local and national taxes and share of employment in order to get a free pass on many regulatory issues.

There is indeed evidence that both regulation and monitoring have had serious shortcomings. Of the several instances of governance failure, I would emphasize two. First, almost all authorities, ranging from the local to the national, that potentially could have had a say in the operation and the monitoring of the lagoons, when interviewed by the Parliamentary Committee investigating the case, claimed a lack of authority. This impotence resulted from the uncertainty around the categorization of red mud: that is, whether it is waste, whether it is toxic, whether it is hazardous, and whether this material belongs to the mining or environmental authorities, as mentioned above.

Another factor is that it was not clear who was responsible for checking the walls' integrity, the composition and toxicity of the red mud, the production process, and the contamination of soil and ground water. It is telling that just two weeks before the catastrophe, the on-site examination of the lagoons by the regional environmental inspectorate found everything in order—an inspection that however did not extend to examining the structural integrity of the pond. Ironically, when the Socialists were in power between 2006–2009, MAL even received the "For Our Environment" award from the Environmental Ministry.

Second, the mechanisms by which the national-level regulatory agencies could have determined whether the nonhazardous categorization of the alumina by-product was correct, or which environmental conditionalities of the privatization could have been enforced, were absent. This too points to a certain lack of coordination among various authorities.

While state ownership in itself is no guarantee of better environmental performance, as the track record of state socialism indicates, it is generally true that state-owned enterprises have what János Kornai has called a soft budget constraint.[21] In practice, that means that the profit motive can be subordinated to other objectives, such as maintaining state control in an industry of strategic or security significance, providing lower prices to consumers, and higher employment; in fact, all three of these are usually quoted as the main rationale for keeping utility companies in state ownership.

In the red mud case, the strong profit motive had two key effects. First, as has now been proven, MAL neglected to realize the environmental remediation, whether prescribed in the privatization contracts or required by environmental regulations, such as the lowering of the pH levels of the sludge to nonhazardous levels, or the closure and recultivation of pond no. 10 by 2010. According to Zoltán Illés, the undersecretary for environment, MAL was considering the closure of the alumina plant within the next few years, perhaps further encouraged in this plan by the economic crisis.[22] Illés cited the absence of any documentation to prove that MAL had been preparing to renew its environmental permit and switch to the dry technology.[23] He also suggested that after filling the ponds to their maximum capacity, which according to some had already been reached by the middle of 2010, they would have closed the plant, leaving debts and environmental damage behind. In a way, this "get-rich-and-get-out-quick" scenario would have been an extended version of the partial privatization strategies mentioned above, that stick the taxpayers with bills.

Second, circumstantial evidence suggests that MAL may have intended to turn the red mud impoundments themselves into profit-making units. This is what I turn to next.

Red Mud as a Source of Profit

Two possibilities of making money out of red mud were available: renting out lagoon space for other wastes, and future recovery of precious raw materials. As we will now see, both projects have their own material conditions that go directly against remediation.

Lagoon Rental

Local residents I interviewed said they knew that MAL accepted other industrial waste for storing in the lagoons. They usually had no direct knowledge of this practice themselves, but quoted acquaintances who had worked at MAL. Such activity certainly would have required a host of environmental permits, and since the postdisaster investigations found no such documentation, MAL, if indeed it engaged in this practice, did so illegally. Therefore, getting people with first-hand knowledge of this to talk to me proved impossible, albeit understandably so. There are two pieces of evidence that MAL may indeed have "rented" out its disposal site to other companies. One comes from the chemical analysis of the spilled sludge. The international environmental organization Greenpeace's laboratory tests found chrome, mercury, and arsenic, compounds that should not have been there, or at least not in such concentrations, had the sludge only been composed of red mud, which is why the Hungarian Academy of Sciences team initially did not even think to look for them.[24] Greenpeace had the tests performed by the Austrian Federal Environmental Protection Agency, on samples collected on the fourth day after the disaster, and then had the results checked by a Hungarian private laboratory that receives commissions from the Hungarian Academy of Sciences itself, which should dispel suspicions of the lab's partiality. Eventually, more than two weeks after the catastrophe, the Academy's own measurements basically confirmed Greenpeace's results.[25] The Greenpeace representative I interviewed interpreted these test results as evidence of illegal dumping of outside wastes.

The second piece of evidence comes from an almost cursory remark in the Parliamentary Committee's report. The document refers to a practice used in the already full and partially recultivated lagoons in Almásfüzitő, whereby hazardous wastes (from outside sources), mixed with organic wastes, were spread on the already full lagoons. The rationale for this procedure was that a relatively impervious layer had to be laid on top of the red mud before soil for plant covering could be spread on the surface, in order to prevent the salination of that soil.[26] According to the committee, however, a clay layer would have served this purpose sufficiently, implying that this was simply an excuse for using the ponds for disposing outside wastes. The committee actually went as far as questioning the premise of composting hazardous by-products by simply mixing them with organic wastes. It is possible that some of the locals had witnessed a similar practice.

Future Recovery of Precious Minerals

Profit motive, however, may have had yet another significant impact on the fate of the alumina waste. According to some informants I talked to, MAL had received offers from Russian and German companies to dredge up and rid the company of the sludge for free. Their interest was in recovering titanium and rare earth minerals that are currently among the most expensive raw materials in the world, or subject to precipitous price rises due to environmental and political considerations. This would have meant a big step toward environmental remediation, because as the technical literature on red mud–disposal methods shows, the greatest difficulty in recultivating such impoundments comes from the need to move the massive volume of sludge safely. MAL, however, allegedly refused these offers because it hoped to recover those minerals itself in the future for its own gain. While right now the recuperation of aluminum, iron, titanium, and rare earth minerals is still very expensive, the burgeoning literature on ever more such recycling methods suggests that it might become economical in the future.[27]

The technical literature—ever more optimistic about not just recuperating aluminum, titanium, and rare earth minerals from red mud but also producing new products from it such as bricks, ground cover, and fertilizer—seems to open up a different trajectory from that of the exhaustion of bauxite resources: not of disappearing, but of newly appearing opportunities. While we have no proof that this is what MAL's owners indeed intended to do, or that they ever would have had the required technology and resources to realize a successful red mud–recovery project, we have various pieces of circumstantial evidence that this scenario has been circulating among the local public. First, Szirmai's 1988 report already refers to the imperative of finding new uses for red mud; as I have shown in my book, reuse enjoyed a high priority in state socialism, and until the early 1980s, a higher one than safe disposal. Second, many of the authors of international technical literature on reuse possibilities have been Hungarian and, given the smallness of the country, it is reasonable to assume that these academic experts and industry representatives formed an epistemic community.[28] In fact, Bakonyi himself had been in charge of Hungary's 1981 Waste and Secondary Raw Material Management Program, which had as its key goal to reduce waste and to find reuses for waste by technological innovation.[29] This makes it reasonable to conclude that he was well-informed about the economic opportunities of recovering expensive materials from red mud, and about the state of the waste management industry. Third, even the 2005 report of the National Development Agency (*Nemzeti Fejlesztési Ügynökség*) on the Ajka small region suggests that the "accumulated red mud could become the raw material for new [economic] activities."[30] In sum, the reuse and recycling of red mud had been talked about enough for these different reports to mention it as a rational and even expected future step.

The significance of this vision for mining the red mud ponds for economic gain resides in the technical synergies between the steps of remediation and mineral recovery. The literature suggests that the different recovery methods on the one hand, and the dewatering and neutralizing of red mud on the other, are parallel and intertwined processes. If I am correct in this interpretation of the technical scholarship, we may have thus gained another explanation for MAL's avoidance of any neutralization and remediation until they were ready to do it themselves, as part of a project of recovering precious minerals.

There are two ways to interpret this attitude to recoverables: ecological modernization or waste-dependent development. In my conclusion, I turn to the question of which interpretation fits the case better.

Conclusion

Ecological modernization is for some a descriptive, while for others a normative theory that emerged in the 1980s. Its proponents have mostly come from Western Europe and Japan, and to a lesser extent from North America. In its most classic version, ecological modernization describes a progressive transition in environmental discourse and practice in the most developed countries.[31] This change consists of a new synergy between economy and ecology, in which economic development (which usually is a synonym for growth) is increasingly beneficial for the environment, rather than, as was the case for much of capitalist history, detrimental. One key element in this relationship is technological innovation, that now allows the prevention, rather than just the treatment, of industrial pollutants, whether air emissions or solid and liquid wastes. This movement away from end-of-pipe solutions, of which smokestacks and tailing dams are the paradigmatic illustrations, is now seen as consistent with efforts to increase efficiency, as inducing further technological innovation, and thus as entirely consistent with the profit motive. Structural transformations accompanying such a trend include cooperative policy making by states and corporations, and collaboration between corporations and nongovernmental organizations in greening production.

"Waste-dependent development" is my term to describe the phenomenon Emily Brownell calls the "new economic order of waste."[32] The key advantage of her approach is that it addresses the blind spots of ecological modernization: primarily global economic relations and social inequality. As some, mostly US-based environmental sociologists have argued, achievements hailed as evidence of ecological modernization in the Global North are made possible by passing down the dirtiest industries to the Global South, where looser environmental standards and enforcement allow these industries to produce more profitably. The alleged decoupling between growth and environmental impact noted by ecological modernization advocates therefore is not so much about greening industry as about leaving the dirtiest for the world's poorest. For waste, it has similarly been argued that the abandonment of end-of-pipe technologies—which in the case of waste primarily refers to dumps and incinerators—in the North has been made possible by sending by-products and discarded consumer goods for recycling to the South, where disassembly, recovery, and recycling are done without the safety and health precautions prescribed by strict northern regulations. Brownell argues that, facilitated by various free trade agreements and waste-debt swaps, developing countries have been increasingly compelled to accept the waste products of the North for recycling or for dumping. It is a well-documented fact that the global waste trade has increased considerably during precisely the period in which the most developed countries allegedly made their production greener, and that key destinations of such trade routes are the former colonies of the world.[33] Relying on resource recovery, recycling, reuse, dumping, incineration, and/or the acceptance of industries with the highest waste ratios and the most toxic by-products as a way to join the world economy or as a way to increase economic activity can indeed be seen as constituting a distinct model of development.

Going even deeper, I suggest that there is also a bifurcation of development mechanisms, in which the richest countries continue to rely primarily on virgin or new raw materials, while the poorest must do with handed-down or disposed-of resources. This is not simply a quantitative difference in the ratio of waste materials in value generation, but a qualitatively different waste-society relationship. Most developed economies rely on the metabolic cycle of new to old, in which value begets waste. But developing economies do not simply rely on the opposite metabolic cycle of old to new, in which waste begets value, but rather one in which waste begets waste. While some value is generated by waste-intensive industries through resource recovery, recycling, reuse, dumping, and incineration, two new forms of waste are also generated: waste in the sense of environmental pollutants, and waste in the sense of squandered resources.

Thus it is clear that waste reuse and recycling are, in theory, key components of both types of development models. How should one evaluate MAL's strategy to delay clean-up, to enable either future reuse or for-profit dumping? Does it exemplify the progressive ecological modernization path, or the altogether more regressive waste-dependent development path that the Global South is following? First of all, one has to answer the question of whether material recovery counts as a preventative method, when the material to be reused has already been dumped and has already wreaked environmental havoc (even prior to the red mud spill). Ecological modernization seems to call for keeping potentially polluting wastes out of nature, which was clearly neither the case nor the intention of the Hungarian alumina company. The use of the red mud lagoons for other industrial waste also puts this case more in the category of classic end-of-pipe technologies than preventative ones.

Therefore, collecting rent from other companies dumping their industrial wastes into the red mud lagoons indicates a path of waste-dependent development, rather than ecological modernization. At the same time, we must modify the above-developed concept of waste-dependent development to make it applicable to Hungary's red mud disaster case. First, MAL stored domestic wastes, rather than, as in the case of the Global South, those of more developed countries. Second, MAL continued to use new raw materials exclusively, and with the exception of aluminum, which is also present in significant amounts in the sludge, most of the materials to be recovered would not have been used in alumina or aluminum production.

MAL did not fit either model of the waste economy. While it may have planned to reuse the by-products, mineral recovery had not been implemented, so ecological modernization clearly does not apply. The promise of future profitable recycling as a key source of company income could certainly have been a motivation in delaying remediation, but this waste-dependent development would not have benefitted a foreign entity, so that model did not fit either.

The best way to capture what path MAL and the Hungarian state had taken, which ultimately led to the disaster, is the adoption of Kevin Hetherington's notion of the gap.[34] In his manifesto for a sociology of disposal—or, more narrowly, for the sociology of consumption to include waste—he argues that wastes are often not truly disposed of, in the sense of settling the matter or the meaning of the discarded materials once and for all. If a thing is thrown away sooner than necessary, the prodigal discarding of value will come back to haunt us as loss or guilt. Alternatively, if we hold onto it too long, it will insinuate itself into our world with its smell and sight. In both cases, a badly managed absence unexpectedly becomes present. Hetherington suggests that such mismanagement is more common than we assume, and one way humans and collectivities have tried to avoid the consequences of such mismanagement is to accomplish disposal in two stages, known from anthropology as a first and a second burial. That is, we need a gap—a transitional, liminal space, if you will—where things can be held and denied the status of wastes, where some value might still be extracted from them.

Hetherington does not concern himself with industrial waste, but nevertheless we can apply his concept of gap. In the case of industrial by-products, it is much more difficult for a producer to create such gaps without violating existing environmental and public health regulations. Yet we can appreciate that, especially in conditions of economic uncertainty and poverty, a company may wish to postpone the final settlement of the discarded matter. In fact, that is exactly what studies focusing on the poor (in the case of consumer waste) suggest. If the regulatory system is weak or in transition, as was the case with MAL, then the company will be able to adopt this gap strategy with impunity. Indeed, as mentioned above, throughout the parliamentary investigation, MAL insisted that red mud is not waste, but rather an intermediary material of alumina processing. Similarly, it wanted to avoid the classification of its ponds as disposal sites, with the argument that materials also leave the lagoons, when some sodium hydroxide is recaptured and returned to the production process. Of course, categorizing the ponds as "technological facilities" (*technológiai berendezések*) rather than waste dumps exempted them from disposal regulations, but it also kept a door open—using Hetherington's metaphor—to access the material for mineral recovery or other reuse technologies. Having complied with disposal regulations and the privatization contract's stipulation that pond no. 10 should be closed by 2010 would have made any future access much more difficult, both from a technological and a legal-regulatory perspective.

The concept of gap is a simple aid in understanding how a postsocialist company amidst radical transformations in property structure, industrial restructuring, globalization, and regulatory reform tried to hedge its bets and played the neither-here-nor-there game that many in those tumultuous years of the 1990s played. Understanding MAL's strategy and rational economic action is not the same as excusing its managers, nor a call for exculpating them. However, such an understanding that pays equal attention to materiality, temporality, and political economy, and to their interaction, could serve as a useful tool for fashioning environmental policy and waste regulation that fits local and non-Western realities better. Let's do it before it is too late: another red mud dump practically identical to Kolontár's, this time in India, is said to face a similar fate.[35]

Zsuzsa Gille is associate professor of sociology at the University of Illinois at Urbana-Champaign. She is author of *From the Cult of Waste to the Trash Heap of History: The Politics of Waste in Socialist and Postsocialist Hungary* (2007), which received the honorable mention of the AAASS Davis Prize, and *Of Paprika, Foie Gras, and Red Mud: The Politics of Materiality in the European Union* (forthcoming).

Notes

1. Taking municipal solid waste as the best proxy for consumer waste, France's rate is 3.5 percent (ADEME 2009). In 2010 in the EU-27, household wastes—another proxy for consumer wastes—was 10 percent (Eurostat 2014).

2. This is probably due to the fact that the liquid content, which is what spilled, contains more NaOH than the settled, more solid sediment at the bottom of the impoundment.
3. Jones, B. E. H., and R. J. Haynes, "Bauxite Processing Residue: A Critical Review of Its Formation, Properties, Storage, and Revegetation," *Critical Reviews in Environmental Science and Technology* 41 no. 3 (2011): 271–315.
4. Greg Power, Markus Gräfe, and Craig Klauber, "Review of Current Bauxite Residue Management, Disposal and Storage: Practices, Engineering and Science," *CSIRO Document DMR3608* (2009): iv, http://www.asiapacificpartnership.org/pdf/Projects/Aluminium/Review%2520of%2520Current%2520Bauxite%2520Residue%2520Man agement%2520Disposal%2520Storage_Aug09_sec.pdf (accessed 12 February 2012).
5. György Bánvölgyi and Tran Minh Huan, "De-watering, Disposal and Utilization of Red Mud: State of the Art and Emerging Technologies," http://www.redmud.org/Files/banvolgyi040110.pdf (accessed 19 January 2012); Robin De Bois, "Red Mud in Hungary: A Predictable, International and Major Disaster," http://www.robindesbois.org/english/risk/red_mud_hungary.html (accessed 19 January 2012); Project Red Mud, "Red Mud Disposal" (n.d.), http://www.redmud.org/Disposal.html (accessed 19 January 2012).
6. A. Juhász, "Development of the Aluminium Industry in Hungary," *Acta Oeconomica* 18, no. 3/4 (1977): 355–369; Ivan T. Berend and György Ránki, *The Hungarian Economy in the Twentieth Century* (London and Sydney, 1985).
7. The economic advantages for Hungary in the various types of economic cooperation and international division of labor were disputed by an analysis of the CIA and, more recently, László Borhi, hinting at a type of economic exploitation characteristic of colonialism. CIA Office of Current Intelligence, "Current Intelligence Bulletin" (Radio Free Europe Archives, 20 June 1952), 5; Borhi, "The Merchants of the Kremlin: The Economic Roots of Soviet Expansion in Hungary," Working Paper No. 28 (Washington, DC, 2000).
8. "Kolontár-jelentés: A vörösiszap-baleset okai és tanulságai" (Kolontár report: Causes and lessons from the red mud disaster), ed. Benedek Jávor and Miklós Hargitai (Budapest, Greens/European Free Alliance Parliamentary Group in the European Parliament and LMP Party, 2011); Johanna Bockman, *Markets in the Name of Socialism: The Left-Wing Origins of Neoliberalism* (Stanford, CA, 2011).
9. "Kolontár-jelentés: A vörösiszap-baleset okai és tanulságai" (Kolontár report: Causes and lessons from the red mud disaster), 4. MAL, in its response to the report of the Parliamentary Committee investigating the disaster, argued that this is a misleading figure, to the extent that the new owner actually also paid some debts of HUNGALU (the name of the company prior to privatization), which amounted to about 433 million forints. This still means that the new owners paid less than one four-hundredth of the nominal value of the company's capital stock. MAL however points out that international accounting firm KPMG valued the company as being worth negative (!) 700 million forints.
10. Of all wastes generated in production, distribution, and consumption, including agricultural waste (which is for the most part recovered in agriculture, and is mainly of plant origin), less than 5 percent is hazardous waste, amounting to 3.4 million tons annually. National Waste Management Plan (2003–2008), http://www.kvvm.hu/szakmai/hulladekgazd/oht_ang.htm (accessed 27 January 2012).

11. Although at that point the dry stacking process was just being implemented in Western Europe, in Hungary a more rudimentary version, in which the red mud was rendered dryer without the sophisticated procedures implemented later in the West, had been practiced since 1972.
12. Viktória Szirmai and Zsuzsa Lehocki, "Környezetállapot es érdekviszonyok Ajkan" (State of the environment and relations of interests in Ajka) (Budapest, Department of Sociology, The College of Political Science of the Hungarian Socialist Workers' Party, 1988).
13. Ibid. Szirmai demonstrates that this was due to the fact that the alumina factory failed to cover ponds nos. 6 and 7 with plants, as instructed by environmental authorities.
14. Eventually these ponds were recultivated, which reduced the dust pollution.
15. The reservoirs nearest to the Danube are ponds nos. 4–7 at Almásfüzitő, adjacent to the country's northern border with Slovakia.
16. "Kolontár report: Causes and lessons from the red mud disaster," 10. MAL, in a criticism of this report, objected to the fact that the committee had not asked for testimonies by geotechnics experts who presumably would have confirmed MAL's explanation of the disaster.
17. The Acquis Communautaire is the collection of EU Treaties, laws, regulation and case law that new entrants to the EU had to adopt into their legislation as a condition of their membership.
18. MAL was correct in arguing that Hungarian Academy of Sciences experts have signed on to the nonhazardousness of red mud, but, according to the investigations of the Parliamentary Committee, the samples on which the ruling was issued came from ponds nos. 1–9, many of which had dried to much lower levels of alkalinity or had stored drier residue to begin with.
19. [N.A.], "MAL-vezér: a vörösiszap nem mérgező, persze fürdeni nem kell benne" (Head of MAL: The red mud is not poisonous, of course one wouldn't want to bathe in it), *HVG*, 5 October, http://hvg.hu/itthon/20101005_iszapomles_mal (accessed 10 February 2012).
20. Zsuzsa Gille, *From the Culture of Waste to the Trash Heap of History: The Politics of Waste in Socialist and Postsocialist Hungary* (Bloomington, IN, 2007).
21. János Kornai, *Economics of Shortage* (Amsterdam, 1980).
22. "Illés: a Mal Zrt. már nem akarta tovább fenntartani a tevékenységét" (Illés: MAL did not want to continue its activity), *168 Óra*, 8 March 2011, http://www.168ora.hu/itthon/illes-a-mal-zrt-mar-nem-akarta-tovabb-fenntartani-a-tevekenyseget-71514.html (accessed 9 October 2014).
23. MAL disputes this, however its arguments are not convincing; [N.A.], "MAL: nem a cég tevékenysége okozta a vörösiszap-katasztrófát" (MAL: The company's activities were not the cause of the red mud catastrophy), *ATV,* 27 December 2011, http://atv.hu/belfold/20111227_mal_nem_a_ceg_tevekenysege_okozta_a_vorosiszap_katasztrofat (accessed 9 October 2014).
24. Greenpeace, "Jóval mérgezőbb az iszap, mint a hivatalos tájékoztatás elmondta" (The sludge is much more toxic than suggested by official information), http://greenpeace.hu/kereses/p1/t272.
25. MTA (Magyar Tudományos Akadémia / Hungarian Academy of Sciences), "Tájékoztató a kolontári vörösiszap tározó környezetében végzett vizsgálatokról" (Information

about the tests performed in the vicinity of the Kolontár red mud impoundment), http://mta.hu/mta_hirei/tajekoztato-a-kolontari-vorosiszap-tarozo-kornyezeteben-vegzett-vizsgalatokrol-125761 (accessed 1 February 2012).
26. This salination danger is due to the high sodium content of red mud, which in turn comes from the NaOH liquid used in alumina production, as described above.
27. Bánvölgyi and Minh Huan, "De-watering, Disposal and Utilization of Red Mud"; Fotini Kehagia, "A Successful Pilot Project Demonstrating the Re-use Potential of Bauxite Residue in Embankment Construction," *Resources, Conservation and Recycling* 54, no. 7 (2008): 417–421; R. K. Paramguru, P. C. Rath, and V. N. Misra, "Trends in Red Mud Utilization—A Review," *Mineral Processing and Extractive Metallurgy* 26, no. 1 (2005): 1–29; Power, Gräfe, and Klauber, "Review of Current Bauxite Residue Management, Disposal and Storage: Practices, Engineering and Science."
28. E.g., Bánvölgyi and Minh Huan, "De-watering, Disposal and Utilization of Red Mud"; P. Siklosi, J. Zoeldi, and E. Singhoffer, "Alumina Industry, Case Study No. 2. Report prepared for UNIDO Conference on Ecologically Sustainable Industrial Development" (Copenhagen, 1991).
29. I interviewed Bakonyi in 1996 already to gather information about the history of Hungary's waste policies.
30. Although it adds, "in the meanwhile it reduces the environmental potential." National Development Agency, 15.
31. Arthur P. J. Mol, *The Refinement of Production: Ecological Modernization Theory and the Chemical Industry* (Utrecht, 1995). Gert Spaargaren and Arthur P. J. Mol, "Sociology, Environment, and Modernity: Ecological Modernization as a Theory of Social Change," *Society and Natural Resources* 5, no. 4 (1992): 323–344.
32. Emily Brownell, "Negotiating the New Economic Order of Waste," *Environmental History* 16, no. 2 (2011): 262–289.
33. Francis O. Adeola, "Cross-National Environmental Injustice and Human Rights Issues: A Review of Evidence in the Developing World," *American Behavioral Scientist* 43, no. 4 (2000): 686–706; S. R. Frey, "The International Traffic in Hazardous Wastes," *Journal of Environmental Systems* 23, no. 2 (1994): 165–177; Bill Moyers, *Global Dumping Ground: The International Traffic in Hazardous Waste* (Washington, DC, 1990).
34. Kevin Hetherington, "Secondhandedness: Consumption, Disposal, and Absent Presence," *Environment and Planning D: Society and Space* 22, no. 1 (2004): 157–173.
35. Amnesty International, "India: Toxic Sludge Leak from Vedanta's Red Mud Pond Threatens Rural Communities," http://www.amnesty.org/en/news-and-updates/india-toxic-sludge-leak-vedantas-red-mud-pond-threatens-rural-communities-2011-06-0 (accessed 9 February 2012).

 CHAPTER 10

The Scramble for Digital Waste in Berlin

Djahane Salehabadi

E-waste (or WEEE) is a shorthand term that refers to a wide range of discarded electrical and electronic equipment, including washing machines, microwaves, hairdryers, light bulbs, vacuum cleaners, computers, cell phones, iPods, and televisions. In Germany, as in the rest of the world, e-waste has become a growing concern over the past decade.

The anxiety over e-waste is due, in part, to the fact that it is one of the world's fastest-growing waste streams.[1] A combination of rapid technological development, planned obsolescence, and perceived obsolescence drive the exponential increase in e-waste production. The problem of overproduction is not limited to the developed world. Emerging economies face similar challenges. A United Nations Environment Program (UNEP) and United Nations University (UNU) estimate that in certain developing countries, e-waste production could increase by up to 500 percent over the next decade.[2]

Not only the abundance, but also the toxicity of e-waste poses a problem. Electrical and electronic goods contain noxious compounds ranging from brominated flame retardants to mercury and arsenic in older equipment.[3] These substances can be released during disposal and recycling. NGOs have released numerous reports characterizing the e-waste issue as an impending environmental crisis of epic proportions.[4] The German media has also enthusiastically taken up the issue of e-waste and stressed this waste stream's toxicity,[5] as have numerous English-language newspapers, magazines, and documentaries.[6]

In addition, NGOs and the media have pointed out the uneven distribution of health and environmental costs associated with the disposal of this waste stream. The export of e-waste is routinely characterized as another form of toxic waste dumping. Accompanied by powerful images of smoldering electronic dumps in China, Bangladesh, Pakistan, and Ghana, NGO and news accounts suggest that export displaces the environmental and social costs of the developed world's high-tech lifestyle to some of the most disadvantaged and vulnerable areas.

However, discarded electrical and electronic equipment can also be valuable. Functional equipment is reusable, and thus holds value in secondhand markets. In addition, discarded technologies can be mined for spare parts, copper, and aluminum, as well as precious metals and rare earth elements. As a result, scavengers, municipal waste authorities, waste-handling firms, and mining companies have become increasingly interested in e-waste as a potential source of revenue.

Thus, far from simply being thrown out and forgotten, e-waste is also an increasingly hot commodity. Various actors involved in the second life of discarded electrical and electronic equipment struggle for control of this material stream. I use the term *struggle* deliberately, since the disposal process is anything but straightforward and unproblematic; instead, it is characterized by conflict and contestation. In this chapter, I outline and analyze the complex afterlives of discarded technologies placed on the curb in Berlin, Germany. Who are the actors that handle Berlin's e-waste? How do they access this rubbish and through what methods do they convert it into a valuable commodity? How do these actors interact with each other? Taking e-waste in Berlin as my point of departure, I engage with these questions and propose a number of preliminary insights into the world of e-waste in Germany and the politics of value creation.

Since August 2009, I have been tracing the life of discarded digital technologies originating in Berlin. My ethnographic and archival research has taken me from Germany to Belgium, Denmark, the Netherlands, Switzerland, and Ghana. For this chapter, I draw on over sixty semistructured interviews of approximately ninety minutes with key actors including local, federal, and EU government officials and industry and NGO representatives, as well as waste handlers. Through these interviews, as well as nonparticipant observation at formal and informal e-waste processing sites, I have been able to reconstruct and analyze the formation and operation of formal and informal recycling networks originating in Berlin.

I focused my study in Berlin for a number of reasons. Germany's capital is its most populated city. It also hosts the country's oldest municipal waste management authority, Berliner Stadtreinigung (BSR). In addition, Berlin is home to the country's largest immigrant scavenger population. These scavengers link the city to the rest of the world through e-waste trade networks.

It is important to note that although the point of departure for this study is Berlin, I conceptualize the city as a node within a complex of global social, economic, cultural, and environmental relationships.[7] Disposal of this waste is organized along an intricate and interconnected global division of labor and ecology. In thinking of Berlin as a node constituted by its relationships to other places and, in turn, constitutive of other places, I seek to unsettle the tendency to reify the boundary of the city or the country in my study.

Germany embodies a paradox. The country is regarded as a global leader in e-waste management. Germany's long-standing reputation as a global leader in solid waste management helps explain this reputation.[8] Germany recognized and addressed the e-waste issue in the mid-1990s, through the IT-Altgeräte-Verordnung (ITV), a regulation on the disposal of information technologies, and played a pivotal role in the development of the European Waste Electrical and Electronic Equipment Directive (WEEE Directive)[9] and the Restriction of Hazardous Substances Directive (RoHS Directive),[10] two of the world's most comprehensive and forward-looking e-waste policies. In 2005, Germany transposed the European directives listed above in the form of the Electrical and Electronic Equipment Act (ElektroG).[11] Member states have substantial leeway in how they interpret the European directives into national policy. Germany's national e-waste policy stands out as thorough and comprehensive.

Despite the country's forward-thinking e-waste policies and extensive waste-handling infrastructure, Germany is not only Europe's biggest e-waste producer, but also probably its biggest exporter, though exact numbers are difficult to come by. Some estimate that Germany can only account for 20 to, at best, 50 percent of its electronic and electrical waste each year.[12] While some of this waste ends up in landfills, experts in the field suspect that the majority is exported. A report by Ökopol, a German environmental policy think tank, estimated that up to 216,0000 tons of e-waste flow out of Germany each year. It is essential to note here, however, that it remains unclear exactly how much of this exported e-waste actually ends up in the developing world.[13]

Formal and Informal E-Waste Recycling Chains

In this section, I outline e-waste's path once it has been discarded. I introduce some of the key actors involved in Germany's e-waste networks and describe how they handle and, where applicable, transform discarded electrical and electronic technologies. Following the example of waste engineers, I separate the e-waste chain into three phases: collection, preprocessing, and recovery.[14] Importantly, despite separating the formal and informal networks here for the sake of clarity, I wish to stress that, in practice, the boundary between these two chains is porous. Actors and objects move fluidly between these two networks. Furthermore, the phases of processing frequently overlap, particularly in the informal processing chain.

German waste law is based on the principle of *geteilte Produktverantwortung*, or shared product responsibility. This means that municipal waste management authorities are responsible for the collection of e-waste. In Berlin, the waste authority is called Berliner Stadtreinigung (BSR). The BSR runs sixteen recycling depots in the city where citizens can drop off their unwanted technologies. Moreover, since 2011, the local government has placed "Orange

Boxes" around the city to make it easier for residents to drop off their unwanted small electrical and electronic items.

While the BSR is responsible for collection, the dual principle mandates that manufacturers pay for the environmentally sound recycling of discarded equipment. An independent clearinghouse, the Elektro-Altgeräteregister (EAR), supervises and coordinates this system. Importantly, manufacturers in Germany are permitted to and almost always do subcontract their responsibility to private third parties.[15] In theory, once the collection bins for electronic wastes at the recycling depots have been filled, manufacturers must pick them up. However, because space is limited in Berlin, a semiprivate company, BRAL Reststoff-Bearbeitungs GmbH (BRAL), transports the containers to a depot outside the city. The BSR and BRAL only handle waste from private households. E-waste from companies goes through entirely different channels.[16]

The municipal waste authority hands over collected items to preprocessors. These companies first manually sort and disassemble e-waste. Alternatively, they mechanically break items apart. After the initial disassembly process comes comminution and, finally, mechanical separation.

Technically, the BSR's role in the recycling chain is restricted to collecting and sorting unwanted electrical and electronic equipment from consumers. The original equipment manufacturers (OEMs) frequently hire private firms to take over preprocessing and recovery on their behalf. However, according to German law, all municipal waste authorities have the right to assume ownership of any category of e-waste for at least one year. Selling e-waste components and fractions represents an important source of revenue for the BSR, and thus the waste authority frequently takes advantage of this option. In Berlin, the BSR extends these rights to BRAL. BRAL not only stores and sorts Berlin's e-waste, but also handles selected parts of this waste stream.

Whether the e-waste is handled at BRAL or at another private facility, the next step after collection is to break apart the equipment into glass, plastic, and metal segments. Depending on the circumstances and the composition of the waste stream, this is either done mechanically or manually. At this point, the larger hazardous and valuable components are extracted. The rest is comminuted. Comminution is the process whereby solid objects are broken down into smaller parts through crushing, grinding, or other techniques. Comminution is followed by mechanical separation in which the fractions are further sorted into aluminum, copper, iron, plastics, and glass.

The next step in the e-waste recycling chain is recovery, whereby valuable materials are extracted and processed into raw materials for production. Preprocessing firms either sell the fractions or components to traders or send them directly to companies specialized in recovery. If the preprocessing firm is small, for instance as with BRAL, then it goes through traders, as recovery firms prefer to deal in large volumes. Traders collect e-scrap, components, or fractions from all over the world before shipping them to recovery and reprocessing firms.

E-waste is broken into three fractions that can be recovered: glass, plastic, and metal. I will focus on the metal fraction in this chapter, as it is the most valuable and thus the most fought-after material stream.

A handful of companies in the world have the necessary equipment for e-scrap recycling. Of these recovery firms, some specialize in iron and aluminum. These fractions are treated to remove contaminants such as lead, tin, and copper. However, because of the physical and chemical properties of iron and aluminum, the elimination of contaminants from this waste stream is challenging. Further, it is nearly impossible to extract traces of precious metals from aluminum and iron mixtures. Given these constraints, aluminum and iron fractions are worth less than the copper fraction. The copper fraction, in contrast, is rich in rare earth metals. Moreover, copper's physical and chemical properties make extracting contaminants and valuable metals much easier.

Only a few firms specialize in the recovery of precious metals from e-scrap. These companies have invested heavily in their e-waste processing infrastructure. These investments attest to the growing importance of e-waste as a source of raw materials and the increasing role of these global mining companies in the process of e-waste recovery.

The recycling process is never 100 percent efficient. Every phase in the e-waste processing chain produces effluent. For instance, preprocessing and recovery produce dust. This dust is often toxic, but can also contain valuable metals. Depending on the quality of the captured dust as well as the market price of metals at any given time, the filters used to capture this dust are sent to facilities to be recycled or are landfilled. Further, the recovery process and incineration produce slag. Some of this slag is used in road construction and for backfilling mines. The slag that is too toxic is deposited in a classified landfill or captured in a concentrated cake, which is then stored.

I would like to make three brief observations on the formal recycling chain before continuing with the informal network. First, formal e-waste handling networks are flexible and dynamic. The structure of the chains changes according to the availability of technology, the material properties of the waste, labor costs, and market demand. Often, companies tailor their disassembly to market demand. That is, if the global demand for circuit boards is high enough, processing firms will sometimes forgo mechanical shredding for manual disassembly in order to preserve whole circuit boards. For this reason, my description of the recycling chain should be thought of as a snapshot in time, as the both the chain as a whole and individual steps within it are constantly in flux.

Second, the e-waste recycling chain is globalized. E-waste brokers and dealers in e-scrap operate in the shadows and are difficult to find. As a result, gaining access to this class of traders is challenging. Nonetheless, I spoke to a number of executives who attested to the global and mobile character of the e-waste market. E-waste rarely stays in one country or region. Instead, e-waste brokers move their "goods" from place to place and often either concentrate on different technologies or transform their items in particular ways depending on market conditions.

Finally, the waste processing and recovery sectors have become increasingly consolidated in the past decade. Where preprocessing firms used to be small- and medium-sized, often family-run operations, today a handful of large companies dominate the global recycling sector.

Despite Germany's sophisticated e-waste management infrastructure, as noted above, a large segment of the country's e-waste remains unaccounted for every year. Some of the missing e-waste ends up in the trash or in storage. Gifting represents another channel of divestment. Used equipment is handed down to friends and relatives or to strangers via forums such as eBay. Charities are also popular ways to get rid of unwanted, yet functional, technologies. That which is not thrown away, stored, or given away to friends, relatives, or charities moves through the country's thriving informal e-waste handling network.

The group that I refer to as informal e-waste handlers or scavengers consists of an amalgamation of multiple players. These actors have different backgrounds and overlapping, yet distinct interests. First, there is a group of individuals—primarily from Eastern Europe—that collects used televisions, computers, and cell phones, as well as household appliances outside Berlin's waste depots. They also sometimes access discarded technologies through flyers and by answering ads in newspapers. This group tends to collect one item at a time and generally does not pay for the goods it collects. Its primary market is Eastern Europe.

Another group, comprised of individuals from certain African and Middle Eastern countries focuses more on larger quantities of items for export. Rather than collect one item at a time, these entrepreneurs tend to buy in bulk from secondhand stores, stores that sell returned goods (so called, A/B/C-Ware), and directly from companies and institutions that have recently upgraded their equipment.[17]

The final cluster consists of a diverse group of individuals who specialize in scrap metal. This group is quite diverse and runs the gamut from collectors who have been in the business for a long time and operate complex and highly lucrative operations to recent immigrants who collect just enough copper or steel in one day to cover their basic living expenses. It is important to note that the line between a scavenger and a legitimate businessperson is often quite blurry. Many scrap dealers have tax numbers and report their earnings, a requirement set by scrap metal firms. However, my interviewees explained that they selectively disclosed their earnings to the authorities.

As noted above, some scavengers manually disassemble items and extract valuable components such as copper wires and yolks or spare parts. One often sees the effects of this process on Berlin streets as unwanted televisions left on the curbside are slowly taken apart over the course of a few days. As with the formal recycling chain, whatever waste handlers deem valueless is handed over to the public facilities for final disposal. For the informal waste handlers, this means either bringing the unwanted items to the BSR recycling stations or leaving unwanted items on the streets. While the "dumping" of leftover waste in the formalized e-waste recycling chain is rarely, if ever, mentioned, the informal "dumping" is a frequent topic in German news reports.[18] Like the formal e-waste processing chain, informal preprocessing extends beyond the city. Informal collectors frequently send collected items to be processed abroad.

Most informal recovery occurs in the developing world. Informal recyclers generally focus on copper and some rare earth metals. For instance, at Agbogbloshie market, the largest e-waste processing site in Accra, Ghana—where I conducted field research during the summer of 2008—children from the impoverished northern part of Ghana break apart and then burn discarded electrical and electronic equipment to extract copper wires. The children sell the copper to scrap dealers, who then export the metals to India, China, and sometimes even the United States and Europe. In other areas, such as Guiyu, China, recyclers use acid baths to extract gold, palladium, and other precious metals from circuit boards. Alternately, individuals may focus on stripping copper wires by hand. Key here is that informal recycling occurs all over the globe; different countries receive different equipment, use different techniques to recycle goods, and focus on different materials or components.

The Struggle for E-Waste

Relatively recent reports by market research firms estimate the e-waste recovery and recycling sector will be worth somewhere between 13 and 20.25 billion US dollars in the next few years.[19] The higher potential profits are, the more pronounced the conflict over e-waste is. In Berlin, three sets of actors vie for control over the e-waste stream: informal actors, local municipalities, and a nexus of manufacturers, scrap dealers, and multinational mining companies. In this section, I sketch the relationships and conflicts between these actors.

The BSR has a vested interest in controlling the e-waste stream. As noted above, e-waste is an important source of revenue for the city's municipal waste authority. Many BSR workers I interviewed ranged from being mildly annoyed by to outright hostile toward informal e-waste collectors and scavengers, especially those that milled outside the collection depots. The administrators I spoke to saw these informal collectors as environmental and petty criminals who routinely steal the city's property.

BRAL and their larger multinational scrap processing counterparts have invested substantial capital in mechanical e-waste processing technologies. Their profits are directly connected to the volume of waste they treat. Mining companies have also invested a lot in their e-waste handling capacity. As a result, these two groups are actively trying to stop the informal e-waste handling networks.

Manufacturers have joined forces with preprocessors and mining companies in an effort to stop the informal e-waste sector. The burden of meeting the collection targets stipulated by European and German e-waste laws falls on manufacturers. They predict that they will have to buy e-waste from informal waste handlers to meet increasing collection quotas. In addition, the bad publicity associated with illegal export is a key motivating factor for manufacturers to stop export. Like the BSR administration, the nexus of preprocessors, mining companies, and manufacturers represents scavengers as thieves and criminals. Interestingly, these actors frequently use pictures and rhetoric from Greenpeace and other environmental organizations—pictures and language that underscore the health and environmental damage associated with backyard recycling techniques such as acid baths and cable burning—to stymie or even reverse the flow of e-waste out of Europe and thus secure their access to this potentially valuable material stream.

Though all three agree on the importance of stopping scavengers, the BSR and the recycling-manufacturing industry are also in conflict with each other. The tension hinges on three issues: collection, the extraction of valuable materials, and environmental responsibility. According to the current ElektroG, German municipalities are the only ones who are allowed to collect e-waste. In exchange, as already mentioned, they have the right to take over ownership of any category of e-waste that could represent a source of revenue. Producers claim that this "cherry-picking" renders the collection process more costly for manufacturers as they are left with the least valuable fractions of this waste stream. Moreover, municipalities would like manufacturers to pay them for the collection costs. Currently, collection is financed through specific waste taxes, which are highly unpopular among Berlin residents. However, manufacturers contend that writing "a blank check" to municipalities who have a monopoly on collection is unfair and inefficient. Manufactures argue that if they should cover the cost of collection, then they should also be allowed to hire private firms. This, they argue, would promote a more efficient and thus more cost-effective system. They add that market competition in collection will eventually lead to more environmentally sound e-waste recycling.

The BSR, in turn, argues that e-waste brings a much-needed source of income for the city of Berlin that struggles under ever-increasing budget cuts.

The profits gleaned from e-waste recycling keeps the city's waste taxes low and subsidizes the costs of treating unprofitable waste streams. In addition, one BSR representative I interviewed questioned the ultimate result of privatization. He argued that privatized recycling had a bad track record in Germany; the moment the waste stream's profitability sank, private firms found ways to dispose of the waste in the cheapest and often highly environmentally irresponsible fashion. Ultimately, the municipal authority would be left with the mess.

Informal e-waste handlers, ranging from exporters and scrap metal collectors to informal recyclers, rarely have the means or venue to respond to their attackers. The Eastern European collectors standing outside the BSR stations were aware of their negative public image. They told me that they were frequently the target of hostility and even violence. They were acutely aware that many saw them as thieves and even environmental criminals. However, they pointed out that their activities were legal.

Whereas waste collectors were frequently confronted with their bad reputations, exporters and scrap collectors seemed to be largely unaware or unconcerned with the campaign to stop them. In fact, exporters often insisted that what they did was not illegal. It is difficult, of course, to determine if they were saying this because they were unfamiliar with Germany's laws or if they were just trying to hide the dubious nature of their work from me. Nonetheless, when I asked them about their work, nearly all the informal e-waste handlers stressed that recycling represented their sole means of securing livelihood. Of course, they understood that what they did could potentially be detrimental to their health, but they had no choice.

Conclusion

In sum, the case of e-waste in Berlin illustrates that disposal is a complex social process worthy of social scientific analysis.[20] The story does not end when an unwanted cell phone, television, or washing machine is labeled trash. Rather, another equally compelling social, political, economic, and cultural tale begins once something is initially discarded.

In Berlin, three sets of actors—scavengers, the municipal waste authority, and the recycling and manufacturing industry—vie for control over the city's e-waste stream. This struggle will likely intensify in the coming decades as consumption and disposal of electrical and electronic equipment increases and as the value of e-waste continues to go up due to resource shortages. Where and by whom should e-waste be treated? Who should ultimately carry the responsibility—financial and otherwise—for this complex waste stream? Who should be allowed to create value out of this particular waste stream? What criteria will be used to evaluate competing e-waste processing techniques?[21] Focusing on technological "fixes" such as more efficient collection, recycling, and recovery techniques alone is unlikely yield meaningful, just, and environmentally sound solutions in the long run. To understand and find solutions to Germany's growing e-waste problem, we need to engage with the complex social and political questions outlined above.

Djahane B. Salehabadi is a senior fellow at the Institute of Environmental Social Sciences and Geography at the University of Freiburg. She received her BA in Earth and Atmospheric Sciences from Dartmouth College, and graduated with a PhD in Development Studies from Cornell University in 2014. Her research and teaching interests lie at the intersection of the environment, technology, and development.

Notes

1. S. Schwarzer et al., *E-Waste, the hidden side of IT equipment's manufacturing and use* (Geneva, 2005)
2. Mathias Schluep et al., *From E-Waste to Resources* (Geneva, 2009).
3. Jim Puckett et al., *Exporting Harm: The High-Tech Trashing of Asia* (Seattle, WA, 2002).
4. Madeleine Cobbing, *Toxic Tech: Not in Our Backyard* (Amsterdam, 2008); Eva Leonhardt, *Geregelte Verantwortungslosigkeit? Erfahrungen mit der Produktverantwortung bei Elektro(nik)-Geräten aus Sicht eines Umwelt- und Verbraucherschutzverbandes* (Berlin, 2007); Puckett, *Exporting Harm*.
5. M. Bitala, "Im Höllenfeuer der Hightech-Welt," *Süddeutsche Zeitung* 17 May 2010; Thorsten Denkler, "Exportschlager Elektroschrott," *Die Tageszeitung,* 16 September 2006; Marc Engelhardt, "Computer-Friedhofe in Afrika," *Welt Sichten,* 9 April 2008; Clemens Höges, "Die Kinder von Sodom," Spiegel, 30.11. 2009; F. Reinbold, "Vergiftete Flammen," *Der Tagesspiegel,* 21 October 2008; NZZ, "Giftiges Gold. In Ghana verbrennen Menschen Elektroschrott, um wertvolle Metalle zu gewinnen," *Neue Zürcher Zeitung,* 25 April 2010.
6. Solly Granatstein and Scott Pelley, "The wasteland", 60 Minutes, CBS Broadcasting, 30 August 2008; Kendra Mayfield, "E-Waste: Dark Side of Digital Age," *Wired,* 1 October 2003, http://www.wired.com/science/discoveries/news/2003/01/57151; Bryan Walsh, "Your laptop's dirty little secret," *Time Magazine,* 30 June 2008; Bryan Walsh, "E-Waste Not: How—and why—we should make sure our old cell phones, TVs and PCs get dismantled properly," *Time Magazine,* 8 January 2009; Richard Wray, "Breeding toxins from dead PCs," *The Guardian,* 6 May 2008.
7. Gillian Hart, *Disabling Globalization: Places of Power in Post-Apartheid South Africa* (Berkeley, CA, 2002); Philip McMichael, "Incorporating Comparison Within a World-Historical Perspective: Towards an Alternative Comparative Method," *American Sociological Review* 55, no. 3 (1990): 385–397; Dale Tomich, "Small Islands and Huge Comparisons: Caribbean Plantations, Historical Unevenness and Capitalist Modernity," *Social Science History* 18, no. 3 (1994): 339–358; Terence Hopkins,

"World-Systems Analysis: Methodological Issues," in *Social Change in the Capitalist World-Economy,* ed. Barbara Hockey Kaplan (New York, 1978), 145–158
8. Miranda S. Schreurs, *Environmental Politics in Japan, Germany, and United States* (Cambridge, 2002).
9. The WEEE Directive regulates collection and recycling of electrical and electronic equipment (EEE). Its objective is to minimize the quantity of EEE in the waste stream and to harmonize the disposal of this waste across European nations.
10. The RoHS Directive restricts the use of certain hazardous substances in EEE, specifically, lead, mercury, cadmium, chromium, VI Polybrominated biphenyl (PBB), and Polybrominated diphenylether (PBDE).
11. Gesetz über das Inverkehrbringen, die Rücknahme und die umweltverträgliche Entsorgung von Elektro- und Elektronikgeräten.
12. Jaco Huisman et al., *2008 Review of Directive 2002/96 on Waste Electrical and Electronic Equipment (WEEE), Final Report* (Bonn, 2008).
13. Knut Sander and Stephanie Schilling, *Optimierung der Steuerung und Kontrolle grenzüberschreitender Stoffströme bei Elektroaltgeräten/Elektroschrott* (Hamburg, 2010).
14. Perrine Chancerel et al., "Assessment of Precious Metal Flows During Preprocessing of Waste Electrical and Electronic Equipment," *Journal of Industrial Ecology* 13, no. 5 (2009): 791–810; Otmar Deubzer, *E-Waste Management in Germany* (Tokyo, 2011).
15. For historical reasons, the private firm ALBA is also permitted to collect some of the city's e-waste in its *Gelbe Tonne* bins. Examining the reasons for this exception goes beyond the scope of this chapter.
16. The afterlives of institutional e-waste are equally complex and worthy of analysis. Again, however, describing these channels goes beyond the scope of this chapter.
17. Here we see the overlap between municipal and institutional e-waste disposal networks.
18. Katharina Schmolinga and Thorsten Bottin, "Abfälle im Wald und im Fluss Entsorgt," *Westdeutsche Allgemeine Zeitung,* 23 July 2010.
19. ABI Research, *E-Waste Recovery and Recycling,* 2010; BCC Research, *Electronic Waste Recovery: Global Markets*; Markets and Markets, *Global E-waste Management Market (2011–2016),* August 2011.
20. Nicky Gregson, Alan Metcalfe, and Louise Crewe, "Moving Things Along: The Conduits and Practices Divestment in Consumption," *Transactions of the Institute of British Geographers* 32, no. 2 (2007): 187–200; Gregson, Metcalfe, and Crewe, *Disposal, devaluation and consumerism: Or how and why things come not to matter* (Sheffield, 2005), internal-pdf://Gregson&Metcalfe_2005-2874829824/Gregson&Metcalfe_2005 .pdf.
21. David N. Pellow, *Garbage Wars: The Struggle for Environmental Justice in Chicago* (Cambridge, 2004).

 PART IV

Reflections

 CHAPTER 11

Can History Offer Pathways to Sustainability?

Donald Worster

This volume is subtitled "Histories of Sustainable Practices." Implied in that phrase is the idea that the history of technology can help us understand best practices for achieving sustainability. Recycling the material goods we produce and consume would be one such practice; adopting the bicycle for transportation would be another. But before we choose any practice, we need to be clear about where we want to go. What is this goal called *sustainability*?

I have read many definitions of the word, and of course they vary in precision and content. The most widely quoted definition derives from the World Commission on Environment and Development's *Our Common Future* (popularly known as the Brundtland Report): meeting "the needs of the present without compromising the ability of future generations to meet their own needs." This definition, like others, is based on the idea that we live in a world of material limits and must share our limited resources with both the present and future generations. In other words, sustainability joins awareness of environmental limits with a sense of global and intergenerational justice. Too frequently, however, the nature of these limits and the exact meaning of justice are not spelled out carefully in current discussions. Are the limits a set of specific resource shortages—not enough food, paper, minerals, or energy to go around—or are the limits really a set of vulnerabilities threatening the earth's natural systems, like climate patterns, ecosystems, fresh water, acidity levels of the oceans, and so forth? How we think about limits and whether we think it is population, consumption, or technology that is threatening those limits makes a big difference.[1]

And what is "justice" after all? Whose definition will we use—that of a Catholic or a Buddhist, or a purely secular definition? Is justice to be defined by capitalists or by socialists or some other economic system's point of view? How many generations into the future should we concern ourselves with? What moral obligations does the nonhuman world lay on us?

It is always tempting to find a simple technological fix for such complex moral and scientific issues. The contributors to this book have not fallen into this trap, but we must remind ourselves that it is a constant danger. The idea that all will be well if you separate your waste into ten separate categories and buy a bicycle is not one that we want to echo.

Once we are clear about the meaning of sustainability, we need to ask what role technology can play in achieving that goal, and what limitations and pitfalls technology may present. Before we can decide on whether any specific technology from the past can be valuable today, we need to ask whether the thinking that invented that older technology can be relevant to problems that we face today. Can the bicycle, for example, offer a significant solution to increasing levels of stratospheric ozone or biodiversity loss? Or are the problems we face going to require more than bicycles—perhaps even a revolution in the ways we think about the earth and technology?

To take the one one example in this volume that links practices of cycling and recycling the closest: M. William Steele's chapter on the Japanese experience shows how that nation is trying to bring back an older Edo culture of frugality and simplicity. Wood, cloth, paper, metalwork, and human feces were all carefully recycled in that period, before capitalism and other Western values invaded. Actually, one could say that before the capitalist revolution, frugality was the norm in all countries, Western as well as Eastern, and unregulated greed was widely viewed as one of the deadly sins. Even as late as World War II, there was some residual feeling in countries as diverse as the United States and Poland that in times of national crisis, the individual citizen must cut consumption and make sacrifices. But in general, life under capitalism has worked to erode that older frugality and social ethic and replaced it with the notion that individual consumption is good for society—that it is our personal duty to consume and promote economic growth. Political leaders may call for certain kinds of restraint from citizens—such as recycling our garbage—but they still want more consumption, for consumption means jobs and jobs will raise the living standard for all. So we are told.

Japan, forsaking its self-imposed isolation, has followed the same international trends toward capitalist logic that one finds in Germany, Canada, or China. How does recycling in today's Japan or Germany function or even contribute to the culture of mass consumption? Does recycling rescue us from modern economic thinking, or merely extend the consumer society indefinitely into the future? Is recycling a moral reform movement, or is its hidden role to make growth sustainable and infinite? It is important that we know more about whether recycling at the Japanese household level has a significant impact on resource use in the greater national economy, which continues to cut down Borneo's rainforests, mine ore in many countries, and export finished goods to nations everywhere. Is the recent effort to promote a more traditional Japanese frugality only serving to mask the nation's global ecological impact? Does an ethic of self-sacrifice at the individual level, enforced by laws and social pressure, have any effect on government and industry's drive for more and more economic growth?

In his chapter, Steele points out the irony of making the bicycle an instrument of social responsibility today, when its invention and diffusion have long been driven by a highly individualistic desire for speed, freedom, and status. The bicycle was invented for the same reasons as the automobile and in some ways, so it seems to me as a pedestrian on the streets of Munich, has engendered even more irresponsible behavior than the automobile has. In its heyday, bicycling was a key expression of the leisured and consumer society.

It is doubtful that bicycles will ever replace automobiles—mostly they add to the number of wheels on the road. Despite the revival of urban bicycles, automobile traffic jams have become monumental in scale. Given a choice, people will buy an automobile, one as large as they can get, and then use the bicycle, as Americans do, for weekend exercise when the weather is right. Would a world of 10 billion bicycles be better for the planet than one of 10 billion automobiles? Or would we end up with 20 billion of each? Such a future would be a disaster.

The fundamental issue facing the planet is whether an economic culture based on endless consumption and material growth can prevent itself from upsetting the many vulnerable earth systems on which life and civilization depend. Perhaps it can be done. Perhaps capitalism can continue to flourish, encouraging entrepreneurial innovation while learning to observe the limits of the earth. This is a question that historians cannot answer definitively but one on which we might provide important perspective. We can examine whether, on the basis of past performance and change, modern cultures based on capitalism or on socialism can be made ecologically responsible.

Thus the role of historians of technology might be to help discover not merely sustainable technologies, but also sustainable economies. Historians should be asking what the limits are of technology in solving threats to the biosphere. Can technologies developed in the past be useful today, or were they developed for a world that no longer exists? Would we be better off, for example, focusing on methods of capturing and sequestering carbon rather than on recycling bottles and cans?

My purpose here is simply to articulate a few of the questions that the essays in this book have stimulated. All of the contributions are commendable in that they try to connect the past with the present and future. They ask what we can take from the past that may be useful. But historians also need to remind the public that we cannot return to some golden age of simple technological solutions, if such an age ever existed. A world population of 6 billion, going on 9–10 billion, will not let us do that. The biggest issue is whether that global population, driven by increasing consumer demands, will permit any kind of return to any part of the past or instead will force us into a new era of radical technological innovation and revolutionary economic change. And if that turns out to be the case, what kind of economic system and culture will be required to achieve sustainability?

Donald Worster held the Hall Distinguished Professorship Chair in American History at the University of Kansas from 1989 to 2010. More recently he has served as distinguished foreign expert at Renmin University of China. His publications include two prize-winning biographies, *A Passion for Nature: The Life of John Muir* (2008), *A River Running West: The Life of John Wesley Powell* (2001), along with eight other books, including: *Rivers of Empire*, which deals with the development of water resources in the West; *Dust Bowl* (1979), a study of the Southern Plains in the "dirty thirties," and *Nature's Economy* (1994, second edition), which traces the development of ecology from the eighteenth century to the present. His books have been translated into French, Italian, Spanish, Chinese, Swedish, Korean, and Japanese.

Notes

1. World Commission on Environment and Development, *Our Common Future* (Oxford, 1987), 43. See also Ulrich Grober, *Sustainability: A Cultural History*, translated by Ray Cunningham (Totnes, Devonshire, 2010); and Donald Worster, "The Shaky Ground of Sustainable Development," *The Wealth of Nature: Environmental History and the Ecological Imagination* (New York, 1993), 142–155.

 CHAPTER 12

History, Sustainability, and Choice

Robert Friedel

Among many of the possible connections to be found in the contributions to this book is a reminder of the fact that the terms *ecology* and *economics* both come from the same root—the Greek οἶκος, referring to the household. To Aristotle, οἰκονομία was the art of properly managing the household. When, in the 1860s, the German naturalist Ernst Haeckel was seeking a term to tie together what Linnaeus had called "the economy of nature" into a concept more useful for his Darwinian leanings, he chose *oekology* to refer to studies that would link all of the elements affecting the struggle for existence to the idea of order and system in the natural world.[1] The intimate linkage, conceptually as well as etymologically, between keeping order in the human household and community, on the one hand, and in maintaining a sustainable order in the earth's environment, on the other, is a valuable message that comes out of a historical examination of cycling and recycling.

It can be argued that the twentieth century, particularly its latter half, will be looked back upon as a remarkable and anomalous period of human history. The growth in world population, the improvements in nutrition and life expectancy, and the wide spread of material affluence over large sections of the globe were not only unprecedented, but they are almost certainly unrepeatable (which is another way of saying unsustainable). The Great Acceleration, to adopt the term of Will Steffen, Paul Crutzen, and John McNeill, was the result, it has been argued, of the confluence of population growth, technological change, and international institutional arrangements that fostered widespread economic expansion.[2] The stories recounted in this volume are, in many ways, stories about confronting the implications of this anomalous period and attempting to manage consequences, both real and anticipated. The paths to sustainability are, as a number of the authors here make clear, paths leading away from the twentieth century anomaly to more familiar patterns of both thought and action.

Seen in this light, the role of history in making these paths clearer becomes more evident. Not only do we need to better understand the sources of the twentieth century, something that historians have worked for generations to do, but we need to contrast the patterns of twentieth-century life and thought both to those more characteristic of human experience in the past and to those that make more sense in the future. In other words, we can ask what it was about institutions, social and political relations, scientific understanding, technological prowess, and human self-awareness that led people throughout much of the world over the last century to abandon practices and behaviors that were clearly, in retrospect at least, far more easily sustained over both time and space than much of what is now seen as normal. Why this war between modernity and sustainability? This is largely a restatement of the key question that Donald Worster asked in the previous chapter: does the past hold the key to recreating a sustainable future, or does it guide us away from the practices that engendered the modern dilemma—or is history really relevant at all?

The first element of an answer to this question lies in better defining what we mean by *sustainability*. Most crudely, we are talking about maintaining the "economy of nature" of the pre-Haeckel biologists. One of the great virtues, however, of the science of ecology that emerged and grew from Haeckel's call is that it came to reveal the dynamic character of nature's economy. The "steady state" does not characterize the natural world, except for relatively brief periods of time. The constancy of change, however, does not mean that there is not orderliness and patterns, greater and lesser degrees of stability. *Sustainable* can thus be understood to mean relationships and activities that can be comfortably maintained over time with little or no depletion of limited resources, with no disruption of the capacity of the environment to accommodate them into the future. In practice, the concept is much more relative than this definition suggests—activities or behaviors are more or less sustainable, depending on the effects they have on continuing action.

There is a degree to which humans, particularly once they become social and technological animals, have always posed challenges to sustainability. The very fact of being omnivores means that human populations have opportunities to outstrip their food supplies while maintaining themselves in the short term. Their remarkable mobility also means humans pose particular challenges to local environments, as they, more than almost any other animal, have the capacity to wreak havoc in one location and then successfully abandon it for another. These capabilities largely predate technologies, which, once they begin to appear and be applied on any scale, increase tremendously the environmental impacts of humans.

All animals extract resources from their environment. For other animals, as a general rule, such extraction is kept in check as part of nature's economy. As extractions increase, populations increase and eventually diminished food supplies act, as Thomas Robert Malthus outlined more than two centuries ago, to check further population increases and thus resource extraction. Malthus, of course, saw this acting as a kind of iron law, forever limiting human populations and human happiness or both. The power of technology to push the Malthusian mechanism aside, particularly over the centuries since he wrote, has changed profoundly the relationship between humans and nature and is, it can be argued, the core cause of the twentieth century anomaly.

Technology alone does not explain, much less determine, the anomaly. As is usually the case, technology enables actions and choices, but it does not shape them. For most of the twentieth century, the key setting in which these actions and choices operated was the market, and the worldwide (though hardly uncontested) triumph of capitalism may be identified as the other indispensable ingredient in the twentieth century's spectacular departure from historical norms.

But technology and capitalism do not, by themselves, truly explain just why the human trajectory took such an unsustainable tack. The essays here, by focusing on a range of activities and conditions, operating both on small personal scales and on larger social and political stages, help us see some of the missing elements. The activities described in most of these chapters seem so reasonable, so simple and ordinary, and largely so benign and even corrective, that it almost seems more sensible to ask not what has promoted bicycling and materials recycling in the late twentieth century, but what has inhibited them? One answer is suggested in the work of the American philosopher Thomas F. Tierney, who wrote of the central place of "convenience" in the calculus of modern life. Tierney maintains that "the value of technology in modernity is centered on technology's ability to provide conveniences." He traces this linkage of technology, modernity, and convenience at least as far in the past as John Locke's argument that humans are not fated to suffer the "inconveniences" of nature (products largely of the Edenic fall from divine grace) if they can devise means around them. This modern technology has done with spectacular success, and so hunger, disease, noisome toil, and stultifying ignorance progressively give way in the world. Where they do not, these parts of the world are dubbed "underdeveloped," which suggests that their accession to this state of greater convenience (and technological grace?) is merely a matter of time and the clearing away of hurdles.[3]

In contrasting and complementary ways, the stories here of bicycles and recycling illustrate efforts to confront the costs of convenience and the construction and reconstruction of alternatives. Cycling—a form of personal transportation that was seen as largely superseded in the most advanced technological societies, most widely by private automobiles—was resurrected in the 1970s in some parts of the industrialized world as a response to the perceived environmental costs of commitments to the automobile. The first question confronted by the chapters here deals with the reasons the bicycle was so widely rejected in the industrialized West. The French case suggests some of the more subtle dimensions of national perceptions of technical choices, with historical images playing roles alongside economic and social forces. The case also reminds us that perceived convenience is contingent on circumstances—a bicycle is a far greater convenience than an automobile in a world in which fuel supplies are tightly controlled and likely to be commandeered at a moment's notice. But perceived convenience in times of hardship and scarcity may result in negative perceptions when circumstances have changed. The Swedish case described here illustrates a pattern probably quite common in the industrialized world, in which accommodation of increased automobility, particularly in the decades after World War II, resulted in "monuments of unsustainability." These monuments—in this instance, bridges designed to inhibit or even prevent cyclist use—are examples of the path dependency and incumbency of infrastructures. As in so much of the increasingly affluent West, Swedes associated motors and not simple mobility with convenience (and an increasingly alluring prosperity), in paths that locked in choices for decades to come, even when some of the larger costs became more evident.

The larger political dimensions of mobility choices become even more evident in two contrasting case studies—of African cultural appropriation of bicycles and of European and American movements to transform the bicycle into a more widely framed "human-powered vehicle." At one level, the African story is fairly simple—describing what the author calls the "material and cultural modifications" of an object as it enters a novel space. From the first appearance of the bicycle in West Africa, it is an instrument of modernity, which becomes an "African modernity" over time. In nonindustrial societies, the bicycle is an accessible instrument of status, mobility, and control, and one that is promoted for its sustainable qualities. This sustainability, however, has its limits, and it reminds us that all sustainability is bounded by circumstances—physical, economic, social, cultural, and intellectual. The politics of bicycle innovation in the West are in stark contrast to the African situation, but they vary enormously contingent on local conditions and experiences. In the Netherlands, the movement for human-powered vehicles became little more than a niche effort to promote a particular kind of bicycle, largely because the commuter bicycle was already a common and accepted part of modern life. In the United States, on the other hand, the same movement was largely an effort to create a sustainable competitor to the automobile, focusing to a large degree on definitions of convenience built around comfort, safety, and speed (definitions unescapably shaped by nearly universal access to automobiles). Both of these cases, the African and the human-powered vehicle, are studies in persistence and innovation, and they emphasize, in very different ways, the political dimensions of both.

To move from a specific technology with a limited range of artifacts, like cycling, to a broad, ill-defined category of material practices, like recycling, is to move the discussion from the limits of a discrete technology and associated uses to a very wide range of attitudes and actions. Recycling, these essays collectively remind us, is as much a mindset as it is a set of practices. In preindustrial societies, both objects and materials were typically the subject of a certain level of stewardship. Even an object no longer fit for its original purposes would be valued for adaptive reuse, and its material for reuse beyond that. What Georg Stöger calls the "premodern mentality of thrift" led naturally to repair, reuse, and the extraction of value. This view contrasts to a degree with that of Roman Köster, who emphasizes that recycling is rooted in scarcity and poverty. This would seem to be a consequence of modern industrialization, where manufacture rather than craftsmanship characterizes the source of things, and thus diminishes the ideals of stewardship. It is not only the ideals that are changed by industrial production, but also the economics.

Industrial manufacture makes things from materials with such relative ease and speed that the market constantly pushes for more output, and the calculus of cost discourages any accounting for disposal, reuse, or risk. From the red mud that is an unavoidable aspect of alumina production to the hazardous metals and solvents that go into electronics, the risks and dangers of modern manufacture are, as much as possible, made external parts of the calculation of value and cost. This is made all the easier by a global system of supply, manufacture, and distribution that encourages at every turn the reduction of cost by the exploitation of unequal wealth and opportunity. This is largely justified by the notion that, whatever its shortcomings, human needs and desires are more fully and more widely met by this system than by any other. But, as Donald Worster has reminded us, the idea of sustainability incorporates not only sufficiency within limits, but also justice.

History tells us that there is no perfect justice, and so we can infer that there is no perfect means for making and distributing things. We have, however, devised what may be the best possible means for maximizing things, for increasing the variety and numbers of products beyond anything previously imaginable. The virtue of applying the historical lens to this fact and to some of its consequences is that we can thus examine not only what is, but also what choices were made to lead to this. Examining the choices leads us in turn to inspect and better understand the values at work in the institutions, communities, and individuals making these choices.

The arguments developed in this volume suggest the possibilities of even deeper challenges to our conceptions of human history and its relation to the present and the future. By looking at how older technologies are used in a range of cultures and times to meet changing needs and circumstances, we are invited to imagine models of history that depart from the conventional linear Western framework—histories that may more resemble in themselves cycles and recycling rather than conventional trajectories of progress. We have only hints here of what these models might look like, but at least the stories here may evoke further thought about the assumptions that most of us typically bring to our environmental anxieties.

Does the role of the historian go beyond this, however? Nearly thirty years ago, the American historians Richard Neustadt and Ernest May pointed out that historical analogies are central to the way that policy makers, politicians, and others frame the challenges before them and formulate possible solutions.[4] Even if every crisis or problem seems unprecedented, it is in our nature to reach back to experience—as individuals and as communities or nations—for guidance about how to respond. Even when the past appears to be a poor guide to new problems, our perceptions of the past shape our expectations about the consequences of possible actions. Paul Sabin makes this argument explicitly in a 2010 article in *Environmental History*, discussing both climate change and energy supply debates.[5] More broadly, the point was made by William Cronon more than two decades ago in discussing "the uses of environmental history": "All of us change the world around us, and yet different people choose to confront their problems and make their changes in strikingly different ways. The diversity of their experiences, past and present, can serve almost as a laboratory for exploring the multitude of choices we ourselves face."[6]

History is indeed a means of studying and understanding choices. Just as, almost 150 years ago, Ernst Haeckel was seeking an approach to nature and to understanding nature's struggles that would incorporate the complexity and dynamism of the whole world, the approach that he dubbed "ecology," a young English economist outlined his view of the central choice that the complex relation between human capabilities and nature's limits posed. William Stanley Jevons was doing his best to understand both the sources and the prospects of the industrial wealth that was remaking his country and the world. His observation was that the power and substance supplied by Britain's coal seams underlay the economic prosperity all around him. He then sought to understand what consequences this fact might have for the future—to what degree, in other words, was this industrial power and wealth sustainable. Using statistics that turned out to be very inaccurate, Jevons came out with a very alarming prediction. Britain's coal, he suggested, would not last even a century, and well before that the increasing difficulty of extracting it would strip the country of any economic advantage it may once have held. It is easy to see in hindsight the limitations of Jevons's vision of the future—in terms of fuels, in terms of technologies, or in terms of the sources of economic growth. But it is less easy to deny the significance of the stark way in which he outlined the fundamental choice that underlies the question of sustainability: "If we lavishly and boldly push forward in the creation of our riches, both material and intellectual, it is hard to over-estimate the pitch of beneficial influence to which we may attain in the present. But the maintenance of such a position is physically impossible.

We have to make the momentous choice between brief but true greatness and longer continued mediocrity."[7]

We are left ultimately with the question whether this is indeed the choice that history lays before us. Or, armed with a keener understanding of the myriad choices made before us, are we now equipped to help supply alternatives?

Robert Friedel has taught environmental history along with history of technology and science at the University of Maryland since 1984. Prior to this, he was a historian at the Smithsonian Institution and the Institute of Electrical and Electronics Engineers. He is the author of six books and numerous articles and essays, and his primary concerns have been with the dynamics of technology, material culture, and social consequences.

Notes

1. Robert C. Stauffer, "Haeckel, Darwin, and Ecology," *The Quarterly Review of Biology* 32 (1957): 140-141.
2. Will Steffen, Paul Crutzen, and John McNeill, "The Anthropocene: Are Humans Now Overwhelming the Great Forces of Nature?" *Ambio: A Journal of the Human Environment* 36, no. 8 (2007): 614-621; Christian Pfister, "The '1950's Syndrome' and the Transition from a Slow-Going to a Rapid Loss of Sustainability," in *The Turning Points of Environmental History*, ed. Frank Uekötter (Pittsburgh, PA, 2010), 90-118.
3. Thomas F. Tierney, *The Value of Convenience: A Genealogy of Technical Culture* (Albany, NY, 1993), esp. 6, 181-186.
4. Richard Neustadt and Ernest May, *Thinking in Time: the Uses of History for Decision-Makers* (New York, 1986); Neustadt and May's well-known efforts are also discussed in Paul Sabin, "'The Ultimate Environmental Dilemma': Making a Place for Historians in the Climate Change and Energy Debates," *Environmental History* 15 (January 2010): 77-78.
5. Sabin, "The Ultimate Environmental Dilemma," 76-93.
6. William Cronon, "The Uses of Environmental History," *Environmental History Review* 17, no. 3 (Autumn 1993): 20.
7. W. Stanley Jevons, *The Coal Question* (London, 1865), 349.

Selected Bibliography

Abbott, Allan V., and David G. Wilson, eds. *Human-Powered Vehicles*. Champaign, IL, 1995.
Adeola, Francis O. "Cross-National Environmental Injustice and Human Rights Issues: A Review of Evidence in the Developing World." *American Behavioral Scientist* 43, no. 4 (2000): 686–706.
Albert de la Bruhèze, Adri A., and Frank C. A. Veraart. *Fietsverkeer in praktijk en beleid in de twintigste eeuw. Overeenkomsten en verschillen in fietsgebruik in Amsterdam, Eindhoven, Enschede, Zuidoost Limburg, Antwerpen, Manchester, Kopenhagen, Hannover en Basel*. Eindhoven, 1999.
Allerston, Patricia. *The Market in Second-hand Clothes and Furnishings in Venice, c. 1500–c. 1650*. PhD thesis, University of Florence, 1996.
Anastas Paul T., and Julie B. Zimmerman. "Design through the 12 Principles of GreenEngineering." *Environmental Science & Technology* 37, no. 5 (2003): 94–101.
Atkinson, A. B., "Optimal Taxation and the Direct versus Indirect Tax Controversy," *Canadian Journal of Economics* 10, no. 590 (1977).
Ballantine, Richard. *Richards Bicycle Book*. New York, 1972.
Banister, David. "The Sustainable Mobility Paradigm." *Transport Policy* 15, no. 2 (2008): 72–80.
Bánvölgyi, György, and Tran Minh Huan. "De-watering, Disposal and Utilization of Red Mud: State of the Art and Emerging Technologies." http://www.redmud.org/Files/banvolgyi040110.pdf (accessed 19 January 2012).
Barth, Ursula. *Frauen gehen lange Wege. Transportvorgänge von Frauen in ländlichen Regionen Afrikas*. Karlsruhe, 1989.
Barwell, Ian. *Le transport et le village. Conclusions d'une série d'enquêtes-villages et d'etudes de cas réalisées en Afrique*. Washington, DC, 1998.
Bausch, Josef H., et al. *Es herrscht Reinlichkeit und Ordnung hier auf den Straßen. Aus 400 Jahren Geschichte der Stadtreinigung und Abfallentsorgung in Dortmund*. Dortmund, 2005.
Beatley, Timothy. *Green Urbanism: Learning from European Cities*. Washington, DC, 2000.
Beck Kurt. "Die Aneignung der Maschine." In *New Heimat*, ed. Karl-Heinz Kohl, 66–77. New York, 2001.
Beckman, Erik, and Svante Linusson, eds. *1000 meter cykelfält som skakade Stockholm*. Bromma, 2009.
Belastingmuseum Prof.dr. Van der Poel, *Rijwielbelasting en de daarvoor gebuikte merken*. Rotterdam, 1978.
Berchicci, Luca. *The Green Entrepreneur's Challenge: The Influence of Environmental Ambition in New Product Development*. PhD thesis, Delft University of Technology, 2005.

Berend, Ivan T., and György Ránki. *The Hungarian Economy in the Twentieth Century.* London and Sydney, 1985.
Berglund, Helge. *Broar och stadsbild: Anförande av Borgarrådet Helge Berglund vid Stockholms Stadsfullmäktiges sammanträde den 17 september 1956.* Stockholm, 1956.
Bertho Lavenir, Catherine *La roue et le stylo. Commes nous sommes devenus touristes.* Paris, 1999.
Berto, Frank J. "The Great American Bicycle Boom." *Cycle History: Proceedings of the International Cycling History Conferences* 10 (1999): 133–141.
———. *The Dancing Chain. History and Development of the Derailleur Bicycle.* San Francisco, CA, 2010.
Beuving, Joost. "Cotonou's Klondike: African Traders and Second-hand Car Markets in Benin." *Journal of Modern African Studies* no. 42 (2004): 511–537.
Bijker, Wiebe E. *Of Bicycles, Bakelites, and Bulbs: Toward a Theory of Sociotechnical Change.* Cambridge, 1995.
Blumberg, Louis, et al. *War on Waste: Can America Win Its Battle with Garbage?* Washington, DC, 1989.
Bockman, Johanna. *Markets in the Name of Socialism: The Left-Wing Origins of Neoliberalism.* Stanford, CA, 2011.
Börjesson, Maria, and Jonas Eliasson. "The Benefits of Cycling: Viewing Cyclists as Travellers rather than Non-Motorists." In *Cycling and Sustainability,* ed. John Parkin, 247–268. London, 2012.
———. "The Value of Time and External Benefits in Bicycle Appraisal." *Transportation Research Part A: Policy and Practice* 46, no. 4 (2012): 673–683.
Boge, Knut. *Votes Count but the Number of Seats Decides: A Comparative Historical Case Study of 20th Century Danish, Swedish and Norwegian Road Policy.* Oslo, 2006.
Borhi, Laszlo G. "The Merchants of the Kremlin: The Economic Roots of Soviet Expansion in Hungary. Working Paper No. 28." Washington, DC, 2000.
Borowy, Iris. "Road Traffic Injuries: Social Change and Development." *Medical History* 57, no. 1 (2013): 108–138.
Bosworth, Richard J. B. "The Touring Club Italiano and the Nationalization of the Italian Bourgeoisie." *European History Quarterly* 27, no. 3 (1997): 371–410.
Bowden, Gregory H. *The Story of the Raleigh Cycle.* London, 1975.
Briese, Volker. "From Cycling Lanes to Compulsory Bike Path: Bicycle Path Construction in Germany, 1897–1940." In *Proceedings of the 5th International Cycle History Conference,* 123–128. 1994.
Brownell, Emily. "Negotiating the New Economic Order of Waste." *Environmental History* 16, no. 2 (2011): 262–289.
Caracciolo, Carlos H. "Bicicleta, circulación vial y espacio público en la Italia Fascista." *Historia Critica* 39 (2009): 20–42.
Carlowitz, Carl von. *Sylvicultura oeconomica oder Haußwirthliche Nachricht und Naturmäßige Anweisung zur Wilden Baum-Zucht,* ed. Joachim Hamberger. Munich, 2013 (1713).
Carlsson, Chris. *Critical Mass: Bicycling's Defiant Celebration.* Oakland, CA, and Edinburgh, 2002.
Carrabine, Eamonn, and Brian Longhurst. "Consuming the Car: Anticipation, Use and Meaning in Contemporary Youth Culture." *The Sociological Review* 50, no. 2 (2002): 181–196.

Cavallès, Robert. "La taxe sur les vélocipèdes." PhD thesis, Université de Toulouse, 1908.
Chakrabarty, Dipesh. "The Climate of History: Four Theses." *Critical Inquiry* 35 (2009): 197–222.
Chancerel, Perrine, et. al. "Assessment of Precious Metal Flows During Preprocessing of Waste Electrical and Electronic Equipment." *Journal of Industrial Ecology* 13, no. 5 (2009): 791–810.
Cobbing, Madeleine. *Toxic Tech: Not in Our Backyard.* Amsterdam, 2008.Cooper, Timothy. "War on Waste: The Politics of Waste and Recycling in Post-war Britain, 1950–1975." *Capitalism, Nature, Socialism* 20, no. 4 (2009): 53–72.
Cordell, H. Ken. "Outdoor Recreation Participation Trends." In *Outdoor Recreation in American Life: A National Assessment of Demand and Supply Trends,* ed. H. Ken Cordell et al., 219–321. Champaign, IL, 1999.
Coulaud, Daniel. *Ville, automobile et modes de vie.* Paris, 2010.
Cox, Peter. *Moving People: Sustainable Transport Development.* London, 2008.
———. "The Role of Human Powered Vehicles in Sustainable Mobility." *Built Environment* 34, no. 2 (2008): 140–160.
Cronon, William. "The Uses of Environmental History." *Environmental History Review* 17, no. 3 (1993): 1–22.
De Bois, Robin. "Red Mud in Hungary: A Predictable, International and Major Disaster." http://www.robindesbois.org/english/risk/red_mud_hungary.html (accessed 19 January 2012).
Deceulaer, Harald. "Second-hand Dealers in the Early Modern Low Countries: Institutions, Markets and Practices." In *Alternative Exchanges: Second-hand Circulations from the Sixteenth Century to the Present,* ed. Laurence Fontaine, 13–42. New York and Oxford, 2008.
Dennis, Richard. *Cities in Modernity: Representations and Productions of Metropolitan Space, 1840–1930.* Cambridge, 2008.
Desforges, Regine. *La bicyclette bleue.* Paris, 1981.
Deubzer, Otmar. *E-Waste Management in Germany.* Tokyo, 2011.
Dovydėnas, Vytas. *Velomobile.* Berlin, 1990.
Dufwa, Arne. *Stockholms tekniska historia: Trafik, broar, tunnelbanor, gator.* Stockholm, 1985.
Duizer, Bram. *"In het nut van actie moet je geloven": Dertig jaar actievoeren door de Fietsersbond.* Utrecht, 2005.
Ebert, Anne-Katrin. "Nationales Design? Auf der Suche nach dem 'Holland-Rad 1900–1940.'" *Technikgeschichte* 67, no. 3 (2009): 211–231.
———. *Radelnde Nationen. Die Geschichte des Fahrrads in Deutschland und den Niederlanden bis 1940.* Frankfurt, 2010.
———. "When cycling gets political: Building cycling paths in Germany and the Netherlands, 1910–40." *Journal of Transport History* 33, no. 1 (2012): 115–37.
Edgerton, David. *The Shock of the Old: Technology and Global History since 1900.* London, 2006.
Eger, Arthur, et al. "Mislukt: Roulandt ligfiets." In *Productontwerpen,* ed. Arthur Eger et al., 237–239. Utrecht, 2004.
Emanuel, Martin. "Constructing the Cyclist: Ideology and Representations in Urban Traffic Planning in Stockholm, 1930–70." *Journal of Transport History* 33, no. 1 (2011): 67–91.
———. *Trafikslag på undantag: Cykeltrafiken i Stockholm 1930–1980.* Stockholm, 2012.

Engelke, Peter O. *Green City Origins: Democratic Resistance to the Auto-Oriented City in West-Germany, 1960–1990.* PhD thesis, Georgetown University, 2011.

Ericsson, Ulla. "Ökad cykelpendling, men hur? En undersökning om attityder till cykling bland boende i innerstadsnära bostadslägen." Stockholm, 2000.

Fehlau, Gunnar. *Das Liegerad.* Kiel, 1996.

Fläschner, Thomas. "Stahlross auf dem Aussterbe-Etat: Zur Geschichte des Fahrrades und seiner Verdrängung in den 50er Jahren." *Eckstein. Journal für Geschichte* no. 9 (2000): 4–22.

Flonneau, Mathieu. *Parcourir et gérer la rue à l'époque contemporaine, Pouvoirs, pratiques, représentations.* Paris, 2008.

Flonneau, Mathieu, and Antoine Prost. *Paris et l'automobile un siècle de passions.* Paris, 2005.

Fontaine, Laurence. "Die Zirkulation des Gebrauchten im vorindustriellen Europa." *Jahrbuch für Wirtschaftsgeschichte* 45, no. 2 (2004): 83–96.

———, ed. *Alternative Exchanges: Second-hand Circulations from the Sixteenth Century to the Present.* New York and Oxford, 2008.

Forester, John. *Bicycle Transportation.* Cambridge, 1983.

Forsell, Håkan. "Den kalla och varma staden: Stockholm som arena för migration och invandring i samtidshistorien: En introduktion." In *Den kalla och varma staden: Migration och stadsförändringar i Stockholm efter 1970*, ed. Håkan Forsell, 9–27. Stockholm, 2008.

Fraunholz, Uwe. *Motorphobia: Anti-automobiler Protest in Kaiserreich und Weimarer Republik.* Göttingen, 2002.

Frey, S. R. "The International Traffic in Hazardous Wastes." *Journal of Environmental Systems* 23, no. 2 (1994): 165–177.

Frilling, Hildegard, and Olaf Mischer. *Pütt un Pann'n. Geschichte der Hamburger Hausmüllbeseitigung.* Hamburg, 1994.

Furness, Zack. *One Less Car: Bicycling and the Politics of Automobility.* Philadelphia, PA, 2010.

Gandy, Mathew. *Recycling and the Politics of Urban Waste.* London, 1994.

Garçon, Anne-Françoise, ed. *L'Automobile, son monde et ses réseaux.* Rennes, 1998.

Geels, Frank, and Johan W. Schot. "Typology of sociotechnical transition pathways." *Research Policy* 36 (2007): 399–417.

Gerlach, Luther. "Traders on Bicycles: A Study of Entrepreneurship and Culture Change among the Digo and Duruma of Kenya." *Sociologus (N.F.)* 13 (1963): 32–49.

Gewald, Jan-Bart, Sabine Luning, and Klaas van Walraven, ed. *The Speed of Change: Motor Vehicles and People in Africa, 1890–2000.* Leiden, 2009.

———. "People, Mines and Cars; Towards a Revision of Zambian History, 1890–1930." In *The Speed of Change*, ed. Jan-Bart Gewald et al., 21–47. Leiden 2009.

Gille, Zsuzsa. *From the Culture of Waste to the Trash Heap of History: The Politics of Waste in Socialist and Postsocialist Hungary.* Bloomington, IN, 2007.

Godard, Xavier, and H. Ngabmen. "Comme Zemidjans, ou le succès des Taxi-Motos." In *Les transports et la ville en Afrique au sud du Sahara. Le temps de la débrouille et du désordre inventif*, ed. Xavier Godard, 397–406. Paris, 2002.

Gosselink, Jack, W., "Vervoer en Fiscus. Blijft de fiets de fiscale verschoppeling?" *Actieradius: vervoermanagement in de Praktijk* 3, no. 5 (1995): 14–15.

Grapperhaus, Ferdinand H. M. *Over de loden last van het koperen fietsplaatje. De Nederlandse rijwielbelasting 1924–1941.* Deventer, 2005.

Greenpeace. "Jóval mérgezőbb az iszap, mint a hivatalos tájékoztatás elmondta." (The sludge is much more toxic than suggested by official information.) http://greenpeace.hu/keres/p1/t272 (accessed 12 February 2012).

Gregson, Nicky, Alan Metcalfe, and Louise Crewe. *Disposal, devaluation and consumerism: Or how and why things come not to matter.* Sheffield, 2005.

———. "Moving Things Along: The Conduits and Practices of Divestment in Consumption." *Transactions of the Institute of British Geographers* 32, no. 2 (2007): 187–200.

Grober, Ulrich. *Sustainability: A Cultural History.* Totnes, 2012.

———. "Von Freiberg nach Rio—Carlowitz und die Bildung des Begriffs ‚Nachhaltigkeit'." In *Die Erfindung der Nachhaltigkeit. Leben, Werk und Wirkung des Hans Carl von Carlowitz*, ed. Sächsische Carlowitz-Gesellschaft, 13–30. Munich, 2013.

Gross, A. C., Ch. R. Kyle, and D. J. Malewicki. "Aerodynamics of Human-Powered Land Vehicles." *Scientific American* (December 1983): 126–134.

Gulick, Charles A. Jr. "Vienna Taxes since 1918." *Political Science Quarterly* 53, no. 4 (1938): 533–556.

Hadland, Tony, and Hans-Erhard Lessing. *Bicycle Design: An Illustrated History.* Cambridge and London, 2014.

Hahn, Hans P. "Global Goods and the Process of Appropriation." In *Resistance and Expansion: Explorations of Local Vitality in Africa*, ed. Peter Probst and Gerd Spittler, 211–230. Münster, 2004.

Hanley, Susan. *Everyday Things in Premodern Japan: The Hidden Legacy of Material Culture.* Berkeley, CA, 1999.

Herlihy, David. V. *Bicycle: The History.* New Haven, CT, and London, 2004.

Hermann, Bernd, and Christof Mauch, eds. *From Exploitation to Sustainability? Global Perspectives on the History and Future of Resource Depletion.* Stuttgart, 2013.

Hetherington, Kevin. "Secondhandedness: Consumption, Disposal, and Absent Presence." *Environment and Planning D: Society and Space* 22, no. 1 (2004): 157–173.

Heyen-Perschon, Jürgen. *Nichtmotorisierte Transportmittel und ihr Einfluss auf die wirtschaftliche und soziale Entwicklung in ländlichen Räumen Ugandas. Empirische Untersuchung zur Kosten-Nutzen-Effizienz der Fahrradnutzung.* Hamburg, 2002.

———. "Das Fahrrad in Afrika – Luxus oder Notwendigkeit." *Praxis Geographie* 32 (2002): 16–20.

Hicks, William Everett, "Shall We Tax the Human Leg?" *North American Review* CLX (October 1897)

Hochenegg, Adolf W. K. "Radfahrsteuer oder nicht?" Leipzig, 1898.

Högselius, Per, Arne Kaijser, and Erik van der Vleuten. *Europe's Infrastructure Transition: Economy, War, Nature.* London 2015.

Hölzl, Richard. "Historicizing Sustainability: German Scientific Forestry in the Eighteenth and Nineteenth Centuries." *Science as Culture* 19, no. 4 (2010): 431–460.

Hoffmann, Aage. "Arbejderidr æ ttens forhold til socialdemokratiet ca. 1880– ca. 1925." *Arbejderhistorie* 1 (2008): 96–115.

Holt, Richard. "The Bicycle, the Bourgeoisie and the Discovery of Rural France." *British Journal of Sports History* 2 (1985): 127–139.

Hommels, Anique. *Unbuilding Cities: Obduracy in Urban Sociotechnical Change.* Cambridge, 2008.
Horton, Dave. "Environmentalism and the Bicycle." *Environmental Politics* 15, no. 1 (2006): 41–58.
Howe, John. "'Filling the Middle': Uganda's Appropriate Transport Services." *Transport Reviews* 23, no. 2 (2003): 161–176.
Huchting, Friedrich. "Abfallwirtschaft im Dritten Reich." *Technikgeschichte* 48, no. 3 (1981): 252–273.
Hunt, Nancy R. "Bicycles, Birth Certificates, and Clysters: Colonial Objects as Reproductive Debris in Mobutu's Zaire." In *Commodification: Things, Agency, and Identities,* ed. Wim van Binsbergen, 123–141 (Münster, 2005).
Huré, Maxime. "Les réseaux transnationaux du vélo; Gouverner les politiques du vélo en ville. De l'utopie associative à la gestion par des grandes firmes urbaines (1965–2010)." PhD thesis, Université Lyon 2 Lumière, 2013.
Illich, Ivan. *Energy and Equity.* London, 1974.
Imura, Hidefumi, and Miranda A. Schreurs, eds. *Environmental Policy in Japan.* Cheltenham, 2005.
International, Amnesty. "India: Toxic Sludge Leak from Vedanta's Red Mud Pond Threatens Rural Communities." http://www.amnesty.org/en/news-and-updates/india-toxic-sludge-leak-vedantas-red-mud-pond-threatens-rural-communities-2011-06-0 (accessed 9 February 2012).
Jamison, Andrew. *The Making of Green Knowledge: Environmental Politics and Cultural Transformation.* Cambridge, 2001.
Jimu, Igansio M. *Urban Appropriation and Transformation: Bicycle Taxi and Handcart Operators in Mzuzu, Malawi.* Mankon, 2008.
Jordan, Pete. *De Fietsrepubliek.* Amsterdam, 2013: 152 and 157.
Jørgensen, Finn Arne. *Making a Green Machine: The Infrastructure of Beverage Container Recycling.* Baltimore, MD, 2011.
Juhász, A. "Development Of The Aluminium Industry In Hungary." *Acta Oeconomica* 18, no. 3/4 (1977): 355–369.
Kaiser, Wolfram, and Johan W. Schot. *Writing the Rules for Europe: Experts, Cartels, International Organizations.* London, 2014.
Kehagia, Fotini. "A Successful Pilot Project Demonstrating the Re-use Potential of Bauxite Residue in Embankment Construction." *Resources, Conservation and Recycling* 54, no. 7 (2008): 417–421.
Kelly, Charles, and Nick Crane. *Richard's Mountain Bike Book.* Sparkford, 1988.
Kerr, Alex. *Dogs and Demons: Tales from the Dark Side of Japan.* New York, 2002.
Kirby, Peter. *Troubled Natures: Waste, Environment, Japan.* Honolulu, HI, 2010.
Kirk, Andrew. "'Machines of Loving Grace': Alternative Technology, Environment, and the Counterculture." In *Image Nation: The American Counterculture of the 1960s and '70s,* ed. Peter Braunstein and Michael W. Doyle, 353–378. New York and London, 2002.
Klercker, Carl-Henrik af. "Trafik- och parkeringsproblemet i Stockholm." *Svenska Kommunal-Tekniska Föreningens Handlingar* 6 (1952): 1–18.
Knoll, Martin, and Reinhold Reith. *An Environmental History of the Early Modern Period: Experiments and Perspectives.* Vienna, 2014.

"Kolontár-jelentés: A vörösiszap-baleset okai és tanulságai" (Kolontár report: Causes and lessons from the red mud disaster), ed. Benedek Jávor and Miklós Hargitai. Budapest, Greens/European Free Alliance Parliamentary Group in the European Parliament and LMP Party, 2011.

Konings, Piet. "'Bendskin' Drivers in Douala's New Bell Neighbourhood: Masters of the Road and the City." In *Crisis and Creativity: Exploring the Wealth of the African Neighbourhood*, ed. Piet Konings and Dick Foeken, 46–65. Leiden, 2006.

Kornai, János. *Economics of Shortage*. Amsterdam, 1980.

Kreuzer, Bernd. "Verkehrszählungen auf Österreichs Straßen zwischen den Weltkriegen und ihr Beitrag zur Verkehrsgeschichte." *Blätter für Technikgeschichte* 60 (1998): 43–62.

———. "1 Fahrrad = 0,25 PKW-Einheiten: Das Fahrrad im Stadtverkehr zwischen verpaßten Chancen und gewollter Marginalisierung, Pfadabhängigkeiten und Gestaltungsspielräumen." In *Erfahrung der Moderne. Festschrift für Roman Sandgruber zum 60. Geburtstag*, ed. Michael Pammer, Herta Neiß, and Michael John, 465–481. Stuttgart, 2007.

———. "Das Fahrrad in Oberösterreich: drei Unternehmen, ein Nebenprodukt und seine Nutzung seit 1870." In *Technikland Oberösterreich. Wirtschaftliche Entwicklungen und industrielle Gegenwart; Symposium, Linz, 22. und 23. Jänner 2010*, ed. Ute Streitt, 151–166. Linz, 2013.

Kuchenbuch, Ludolf. "Abfall: Eine stichwortgeschichtliche Erkundung." In *Mensch und Umwelt in der Geschichte*, ed. Jörg Calließ et al., 257–276. Pfaffenweiler, 1989.

Kuipers, Giselinde. "The Rise and Decline of National Habitus: Dutch Cycling Culture and the Shaping of National Similarity." *European Journal of Social Theory* 16, no. 1 (2012): 17–35.

Kyle, Chester. "A History of Human-Powered Land Vehicles and Competition." In *Humanpowered Vehicles*, ed. Allan V. Abbott and David G. Wilson, 95–111. Champaign, IL, 1995.

———. "Bicycle Aerodynamics and the Union Cycliste Internationale." *Cycle History: Proceedings of the International Cycling History Conferences* 11 (2000): 118–131.

———. "A Brief History of the International Human-Powered Vehicle Association 1976–1998." *Cycle History: Proceedings of the International Cycling History Conferences* 12 (2001): 134–145.

Lambert, Benoît. *Cyclopolis, ville nouvelle: Contribution à l'histoire de l'écologie politique*. Geneva, 2004.

Lemire, Beverly. "The Theft of Clothes and Popular Consumerism in Early Modern England." *Journal of Social History* 24 (1990): 255–276.

———. "Shifting Currency: The Culture and Economy of the Second Hand Trade in England, c. 1600–1850." In *Old Clothes, New Looks: Second Hand Fashion*, ed. Alexandra Palmer and Hazel Clark, 29–47. Oxford and New York, 2006.

Leonhardt, Eva. *Geregelte Verantwortungslosigkeit? Erfahrungen mit der Produktverantwortung bei Elektro(nik)-Geräten aus Sicht eines Umwelt- und Verbraucherschutzverbandes*. Berlin, 2007.

Linder, Staffan. *Harried Leisure Class*. New York, 1970.

Linders-Rooijendijk, Matea F. A. *Gebaande wegen voor mobiliteit en vrijetijdsbesteding. De ANWB als vrijwillige organisatie 1883–1937*. Heeswijk-Dinther, 1989.

Löfwander, Torild. *Die sozialökonomischen Verhältnisse der bäuerlichen Bevölkerung in der Republik Mali.* Berlin, 1983.
Longhurst, James. "The Sidepath Not Taken: Bicycles, Taxes, and the Rhetoric of the Public-Good in the 1890s." *The Journal of Policy History* 25, no. 4 (2013): 557–586.
———. "'Awheel from Chicago to the Twin Cities': Legacies of Turn-of-the-Century Bicycle Paths in Minneapolis and St. Paul." In *Two Cities, One Hinterland: An Environmental History of the Twin Cities and Greater Minnesota*, ed. George Vrtis and Christopher W. Wells. Forthcoming.
Lukes, Steven. *Maktens ansikten.* Stockholm, 2008.
Lundin, Per. "Att tänka om staden med historia: En introduktion till Bilstaden." In *Bilstaden: USA visade vägen*, ed. Uno Åhrén and Per Lundin, 9–22. Stockholm, 2010 (1960).
Männistö-Funk, Tiina. "The Prime, Decline and Recalling of Rural Cycling: Finnish Cycling in the 1920s and 1930s Remembered in 1971–1972." *Transfers* 2, no. 2 (2012): 49–69.
Mahoney, James. "Path Dependence in Historical Sociology." *Theory and Society* 29, no. 4 (2000): 507–548.
Malmsten, Bo. *Dennisöverenskommelsen: Förhandlingen, aktörerna, innehållet.* Stockholm, 1993.
Mauch, Christof, and Helmuth Trischler. *International Environmental History: Nature as a Cultural Challenge.* Munich, 2010.
McCullagh, James C., ed. *Pedal Power in Work, Leisure, and Transportation.* Emmaus, PA, 1977.
McDonough, William, and Michael Braungart. *Cradle to Cradle: Remaking the Way We Make Things.* New York, 2002.
———. *The Upcycle.* New York, 2013.
McFarlane, Colin, and Jonathan Rutherford. "Political Infrastructures: Governing and Experiencing the Fabric of the City." *International Journal of Urban and Regional Research* 32, no. 2 (2008): 363–374.
Melosi, Martin. *Garbage in the Cities: Refuse, Reform, and the Environment.* Pittsburgh, PA, 2005.
Meyer, Jan-Henrik. "Saving Migrants: A Transnational Network Supporting Supranational Bird Protection Policy." In *Transnational Networks in Regional Integration: Governing Europe, 1945–83*, ed. Wolfram Kaiser, Brigitte Leucht, and Michael Gehler, 176–198. London, 2010.
Ministerie van Verkeer en Waterstaat, *Eindrapport Masterplan Fiets. Samenvatting, evaluatie en overzicht van de projecten in het kader van het Masterplan Fiets, 1990–1997.* Den Haag, 1998.
Möllers, Nina, Christian Schwägerl, and Helmuth Trischler, eds. *Welcome to the Anthropocene: The Earth in Our Hands.* Munich, 2015.
Moghaddass Esfehani, Amir. "The Bicycle's Long Way to China: The Appropriation of Cycling as a Foreign Cultural Technique 1860–1940." In *Cycle History 13. Proceedings, 13th International Cycling History Conference*, ed. Andrew Ritchie and Rob van der Plas, 94–102. San Francisco, CA, 2003.
Mol, Arthur P. J. *The Refinement of Production: Ecological Modernization Theory and the Chemical Industry.* Utrecht, 1985.

Mom, Gijs. "Decentering Highways: European National Road Network Planning from a Transnational Perspective." In *In der Moderne Strasse: Planung, Bau und Verkehr vom 18. bis zum 20. Jahrhundert*, ed. Hans-Liudger Dienel and Hans-Ulrich Schiedt, 77–100. Frankfurt, 2010.

Mom, Gijs, and Ruud Filarski. *Van transport naar mobiliteit. De mobiliteitsexplosie (1895–2005)* Zutphen, 2008.

Moyers, Bill. *Global Dumping Ground: The International Traffic in Hazardous Waste.* Washington, DC, 1990.

MTA (Magyar Tudományos Akadémia / Hungarian Academy of Sciences). "Tájékoztató a kolontári vörösiszap tározó környezetében végzett vizsgálatokról." (Information about the tests performed in the vicinity of the Kolontár red mud impoundment.) http://mta.hu/mta_hirei/tajekoztato-a-kolontari-vorosiszap-tarozo-kornyezeteben-vegzett-vizsgalatokrol-125761 (accessed 1 February 2012).

Münch, Peter. *Stadthygiene im 19. und 20. Jahrhundert. Die Wasserversorgung, Abwasser- und Abfallbeseitigung unter besonderer Berücksichtigung Münchens.* Göttingen, 1993.

[N.A.] "Illés: a Mal Zrt. már nem akarta tovább fenntartani a tevékenységét." (Illés: MAL did not want to continue its activity.) *168 Óra*, 8 March 2011. http://www.168ora.hu/itthon/illes-a-mal-zrt-mar-nem-akarta-tovabb-fenntartani-a-tevekenyseget-71514.html (accessed 9 October 2014).

[N.A.] "MAL: nem a cég tevékenysége okozta a vörösiszap-katasztrófát." (MAL: The company's activities were not the cause of the red mud catastrophy.) *ATV*, 27 December 2011. http://atv.hu/belfold/20111227_mal_nem_a_ceg_tevekenysege_okozta_a_voro siszap_katasztrofat (accessed 9 October 2014).

[N.A.] "MAL-vezér: a vörösiszap nem mérgező, persze fürdeni nem kell benne." (Head of MAL: The red mud is not poisonous, of course one wouldn't want to bathe in it.) *HVG*, 5 October 2010. http://hvg.hu/itthon/20101005_iszapomles_mal (accessed 10 February 2012).

Nast, Matthias. *Die stummen Verkäufer. Lebensmittelverpackungen im Zeitalter der Konsumgesellschaft. Umwelthistorische Untersuchung über die Entwicklung der Warenverpackung und den Wandel der Einkaufsgewohnheiten (1950er bis 1990er Jahre).* Berlin, 1997.

Nilsson, Lars. "The Return to the City: Twentieth Century Urban Development in Sweden." In *Reclaiming the City: Innovation, Culture, Experience*, ed. Marjaana Niemi and Ville Vuolanto, 47–62. Helsingfors, 2003.

Nwabughuogu, Anthony I. "The Role of Bicycle Transport in the Economic Development of Eastern Nigeria, 1930–45." *Journal of Transport History* 5 (1984): 91–98.

Oldenziel, Ruth, and Adri Albert de la Bruhèze. "Contested Spaces: Bicycle Lanes in Urban Europe, 1900-1995." *Transfers* 1, no. 2 (2011): 31–49.

———."Cycling in a Global World. Special Section." *Transfers* 2, no. 2 (2012).

Oldenziel, Ruth, and Mikael Hård. *Consumers, Users, Rebels: The People Who Shaped Europe*. London, 2013.

Oldenziel, Ruth, and Milena Veenis. "The Glass Recycling Container in the Netherlands: Symbol in Times of Shortages and Abundance, 1939–1978." *Contemporary European History* 22, no. 3 (2013): 453–476.

Oldenziel, Ruth, and Heike Weber. "Introduction: Reconsidering Recycling." *Contemporary European History* 22, no. 3 (2013): 347–370.

Paramguru, R. K., P. C. Rath, and V. N. Misra. "Trends in Red Mud Utilization—A Review." *Mineral Processing and Extractive Metallurgy* 26, no. 1 (2005): 1–29.
Park, M. A. Jinhee. *Von der Müllkippe zur Abfallwirtschaft. Die Entwicklung der Hausmüllentsorgung in Berlin (West) von 1945 bis 1990*. Berlin, 2004.
Pellow, David N. *Garbage Wars: The Struggle for Environmental Justice in Chicago*. Cambridge, 2004.
Pfister, Christian, ed. *Das 1950er Syndrom: Der Weg in die Konsumgesellschaft*, 2nd ed. Bern, 1996.
——. "The '1950s Syndrome' and the Transition from a Slow-Going to a Rapid Loss of Global Sustainability." In *The Turning Points in Environmental History*, ed. Frank Uekötter, 90–118. Pittsburgh, PA, 2010.
Pinch, Trevor J., and Wiebe E. Bijker. "The Social Construction of Facts and Artifacts: Or How the Sociology of Science and the Sociology of Technology Might Benefit Each Other." In *The Social Construction of Technological Systems*, ed. Wiebe E. Bijker, Thomas P. Hughes, and Trevor E. Pinch, 17–50. Cambridge, 1987.
Pivato, Stefano. "The bicycle as a political symbol: Italy, 1885–1955." *International Journal of the History of Sport* 7, no. 2 (1990): 172–187.
Polaschek, Martin F. "Funktionierender Parlamentarismus im Ständestaat? Die Auseinandersetzungen um die Einführung einer Fahrradabgabe in der Steiermark." *Zeitschrift des Historischen Vereines für Steiermark* 86 (1995): 277–301.
Pooley, Colin G., and Jean Turnbull. "Modal Choice and Modal Change: The Journey to Work in Britain since 1890." *Journal of Transport Geography* 8, no. 1 (2000): 11–24.
Porter, Gina. "Living in a Walking World: Rural Mobility and Social Equity Issues in Sub-Saharan Africa." *World Development* 30, no. 2 (2002): 285–300.
Power, Greg, Markus Gräfe, and Craig Klauber. "Review of Current Bauxite Residue Management, Disposal and Storage: Practices, Engineering and Science." *CSIRO Document DMR3608* (2009). http://www.asiapacificpartnership.org/pdf/Projects/Aluminium/Review%2520of%2520Current%2520Bauxite%2520Residue%2520Management%2520Disposal%2520Storage_Aug09_sec.pdf (accessed 12 February 2012).
Prynn, David. "The Clarion Clubs, Rambling and the Holiday Associations in Britain since the 1890s." *Journal of Contemporary History* 11, no. 2/3 (1976): 65–77.
Pucher, John. "Bicycling Boom in Germany: A Revival Engineered by Public Policy." *Transportation Quarterly* 51 (1997): 31–46.
Pucher, John, and Ralph Buehler. "Making Cycling Irresistible: Lessons from the Netherlands, Denmark and Germany." *Transport Reviews* 28, no. 4 (2008): 495–528.
Pucher, John, Charles Komanoff, and Paul Schimek. "Bicycling Renaissance in North America? Recent Trends and Alternative Policies to Promote Bicycling." *Transportation Research Part A* 33, no. 7/8 (1999): 625–54.
Puckett, Jim, et al. *Exporting Harm: The High-Tech Trashing of Asia*. Seattle, WA, 2002.
Pye, Denis. *Fellowship is Life: The National Clarion Cycling Club 1895–1995*. Bolton, 1995.
Rabenstein, Rüdiger. *Radsport und Gesellschaft: ihre sozialgeschichtlichen Zusammenhänge in der Zeit van 1867 bis 1914*. Hildesheim, 1955.
Radermacher, Franz Josef. "Die Ressourcen der Erde setzen uns Grenzen – vom sächsischen Bergmann Hans Carl von Carlowitz 1713 bis zum neuen Report an den Club of Rome 2052." In *Die Erfindung der Nachhaltigkeit. Leben, Werk und Wirkung des Hans Carl von Carlowitz*, ed. Sächsische Carlowitz-Gesellschaft, 141–156. Munich, 2013.

Ranger, Terence. "Bicycles and the Social History of Bulawayo." In *Short Writings from Bulawayo*, ed. Jane Morris, 76–81. Bulawayo, 2003.
Red Mud, Project. "Red Mud Disposal." N.d. http://www.redmud.org/Disposal.html (accessed 19 January 2012).
Reith, Reinhold. "Recycling im späten Mittelalter und der frühen Neuzeit: Eine Materialsammlung." *Frühneuzeit-Info* 14 (2003): 47–65.
———. *Umweltgeschichte der Frühen Neuzeit*. Munich, 2011.
Reith, Reinhold, and Georg Stöger. "Western European Recycling in a Long-term Perspective: Reconsidering Caesuras and Continuities." *Jahrbuch für Wirtschaftsgeschichte* 56, no. 1 (2015).
———. "Reparieren – oder die Lebensdauer der Gebrauchsgüter." *Technikgeschichte* 79, no. 3 (2012): 173–184.
Reitsma, S.A. "De rijwielbelasting als bestemmingsheffing voor het motorsnelverkeer." *De Opbouw* 2 (1938).
———. *Herwaardering van Verkeerseconomische Waarden*. Den Haag, 1942.
Renne, Elisha P., and Dakyes S. Usman, "Bicycle Decoration and Everyday Aesthetics in Northern Nigeria." *African Arts* 32, no. 2 (1999): 46–51.
Righart, Hans. *De eindeloze jaren zestig. Geschiedenis van een generatieconflict*. Amsterdam, 1995.
Rome, Adam. *The Bulldozer in the Countryside: Suburban Sprawl and the Rise of American Environmentalism*. Cambridge, 2001.
———. *The Genius of Earth Day: How a 1970 Teach-in Unexpectedly Made the First Green Generation*. New York, 2013.
Rosen, Paul. "Up the Vélorution: Appropriating the Bicycle and the Politics of Technology." In *Appropriating Technology: Vernacular Science and Social Power*, ed. Ron Eglash, et al., 365–389. Minneapolis, MN, 2004.
Rosenberg, Daniel, and Anthony Grafton. *Cartographies of Time: A History of the Timeline*. Princeton, NJ, 2010.
Rosenberg, Daniel, and Susan Harding, eds. *Histories of the Future*. Durham, NC, 2005.
Sabin, Paul. "'The Ultimate Environmental Dilemma': Making a Place for Historians in the Climate Change and Energy Debates." *Environmental History* 15 (2010): 76–93.
Sachs, Wolfgang. *For Love of the Automobile: Looking Back into the History of Our Desires*. Berkeley, CA, 1992.
Sandberg, Brian. "'The Magazine of All Their Pillaging': Armies as Sites of Second-hand Exchange during French Wars of Religion." In *Alternative Exchanges: Second-hand Circulations from the Sixteenth Century to the Present*, ed. Laurence Fontaine, 76–96. New York and Oxford, 2008.
Sander, Knut, and Stephanie Schilling. *Optimierung der Steuerung und Kontrolle grenzüberschreitender Stoffströme bei Elektroaltgeräten/Elektroschrott*. Hamburg, 2010.
Sanderson, Elizabeth C. "Nearly New: The Second-hand Clothing Trade in Eighteenth Century Edinburgh." *Costume* 31 (1997): 38–48.
Sandin, Gunnar. *Vägen till citybanan: Spårfrågan mellan Norr och Söder under 150 år*. Stockholm, 2012.
Schluep, Mathias, et al. *From E-Waste to Resources*. Geneva, 2009.
Schmitz, Arnfried, and Tony Hadland. *Human Power: The Forgotten Energy 1913–1992*. Coventry, 2000.

Schneider, Wolfgang. *Sekundärrohstoff Altpapier. Markt und Marktentwicklung in der Bundesrepublik Deutschland.* Dortmund, 1988.
Schreurs, Miranda S. *Environmental Politics in Japan, Germany, and United States.* Cambridge, 2002.
Schroeder, Bradley. "Doing Business in Africa: The California Bike Coalition Comes of Age." *Sustainable Transport* 18 (2007): 18–21.
Schwartz, Leslie. "Fietsbelasting in Nederland, 1924–1941," *Genealogie* 12, no. 2 (2006): 51–52.
Schwarzer S., et al. *E-Waste, the hidden side of IT equipment's manufacturing and use.* Geneva, 2005.
Scott, James C. *Seeing Like a State: How Certain Schemes to Improve the Human Condition Have Failed.* New Haven, CT, 1998.
Shove, Elizabeth. "The Shadowy Side of Innovation: Unmaking and Sustainability." *Technology Analysis & Strategic Management* 24, no. 4 (2012): 363–375.
Siklosi, P., J. Zoeldi, and E. Singhoffer. *Alumina Industry, Case Study No. 2. Report prepared for UNIDO Conference on Ecologically Sustainable Industrial Development.* Copenhagen, 1991.
Skårfors, Rikard. *Stockholms trafikledsutbyggnad: Förändrade förutsättningar för beslut och implementering 1960–1975.* Uppsala, 2001.
Spaargaren, Gert, and Arthur P. J. Mol. "Sociology, Environment, and Modernity: Ecological Modernization as a Theory of Social Change." *Society and Natural Resources* 5, no. 4 (1992): 323–344.
Spittler, Gerd. "Kleidung statt Essen. Der Übergang von der Subsistenzproduktion zur Marktproduktion bei den Hausa (Niger)." In *Afrika zwischen Subsistenzökonomie und Imperialismus*, ed. Georg Elwert and Roland Fett, 93–105. Frankfurt, 1982.
Staal, Peter E. *Automobilisme in Nederland: Een geschiedenis van gebruikers, misbruik en nut.* Zutphen, 2003.
Stahre, Ulf. *Den gröna staden: Stadsomvandling och stadsmiljörörelse i det nutida Stockholm.* Stockholm, 2004.
———. *Reclaim the streets: Om gatufester, vägmotstånd och rätten till staden.* Stockholm, 2010.
Stauffer, Robert C. "Haeckel, Darwin, and Ecology." *The Quarterly Review of Biology* 32 (1957): 138–144.
Steele, M. William. "The Speedy Feet of the Nation: Bicycles and Everyday Mobility in Modern Japan." *Journal of Transport History* 32, no. 2 (2011): 187–209.
———. "The Making of a Bicycle Nation: Japan." *Transfers* 2, no. 2 (2012): 70–94.
Steffen, Will, et al. "The Anthropocene: Conceptual and Historical Perspectives." *Philosophical Transactions of the Royal Society Academy A* 369 (2011): 842–867.
Steffen, Will, Paul J. Crutzen, and John R. McNeill. "The Anthropocene: Are Humans Now Overwhelming the Great Forces of Nature?" *Ambio* 36 (2007): 614–621.
Stobart, Jon, and Ilja Van Damme, eds. *Modernity and the Second-hand Trade: European Consumption Cultures and Practices, 1700–1900.* Basingstoke, 2010.
Stöger, Georg. *Sekundäre Märkte? Zum Wiener und Salzburger Gebrauchtwarenhandel im 17. und 18. Jahrhundert.* Vienna and Munich, 2011.
———. "Disorderly Practices in the Early Modern Urban Second-hand Trade (Sixteenth to Early Nineteenth Centuries)." In *Shadow Economies and Irregular Work in Urban Eu-*

rope: 16th to Early 20th Centuries, ed. Thomas Buchner and Philip Hoffmann-Rehnitz, 141–163. Berlin, 2011.
Stoffers, Manuel. "The Human Powered Vehicle Movement and the Changing Image of the Bicycle at the End of the Twentieth Century." *Cycle History* 22 (2012): 211–219.
———. "Cycling as Heritage: Representing the History of Cycling in the Netherlands." *Journal of Transport History* 33, no. 1 (2012): 92–114.
Stoffers, Manuel, and Anne-Katrin Ebert. "New Directions in Cycling Research: A Report on the Cycling History Roundtable at T2M Madrid." In *Mobility in History* 5, ed. Peter Norton et al., 9–19. New York and Oxford, 2014.
Stokes, Raymond G., Roman Köster, and Stephen Sambrook. *The Business of Waste: Britain and Germany, 1945 to the Present.* Cambridge, 2013.
Strandberg, Samuel. "Kursändring hur fort?" In *Gang-, cykel- og knallerttrafikken som integreret del af transportudviklingen: Foredrag m.v. ved NKTF:s konference den 16–18 april 1980 i København.* Copenhagen, 1980.
Strasser, Susan. *Waste and Want: A Social History of Trash.* New York, 1999.
Szirmai, Viktória, and Zsuzsa Lehocki. "Környezetállapot es érdekviszonyok Ajkan." (State of the environment and relations of interests in Ajka.) Budapest, Department of Sociology, The College of Political Science of the Hungarian Socialist Workers' Party, 1988.
Tierney, Thomas F. *The Value of Convenience: A Genealogy of Technical Culture.* Albany, NY, 1993.
Tjong Tjin Tai, Sue-Yen, Frank Veraart, and Mila Davids. "How the Netherlands became a bicycle nation: Users, firms and intermediaries, 1860–1940." *Business History* (2014): 1–33.
Trischler, Helmuth, ed. *Anthropocene: Exploring the Future of the Age of Human.* Munich, 2013.
Turner, Fred. *From Counterculture to Cyberculture: Stewart Brand, the Whole Earth Network, and the Rise of Digital Utopianism.* Chicago, IL, 2006.
Ugorji, Rex U., and Nnennaya Achinivu. "The Significance of Bicycles in a Nigerian Village." *The Journal of Social Psychology* 102 (1977): 241–246.
Ui, Jun, ed. *Industrial Pollution in Japan.* New York, 1991.
Verrips, Jojada, and Birgit Meyer. "Kwaku's Car: The Struggles and Stories of a Ghanaian Long-distance Taxi-driver." In *Car Cultures*, ed. Daniel Miller, 153–184. Oxford, 2001.
Van Dam, Cees, and Gerrit van Maanen. "The development of traffic liability in the Netherlands." In *The Development of Traffic Liability*, ed. Wolfgang Ernst, 112–150. Cambridge, 2010.
Van Damme, Ilja, and Reinoud Vermoesen. "Second-Hand Consumption as a Way of Life: Public Auctions in the Surroundings of Alost in the Late Eighteenth Century." *Continuity and Change* 24, no. 2 (2009): 275–305.
Veraart, Frank C. A. "Geschiedenis van de fiets in Nederland, 1870–1940." MA Thesis, TU Eindhoven, 1995.
Weber, Donald. *De blijde intrede van de automobile in België, 1895–1940.* Gent, 2010.
Wehap, Wolfgang. *Frisch, Radln, Steierisch: Eine Zeitreise durch die regionale Kulturgeschichte des Radfahrens.* Graz, 2005.
Welleman, Ton. *Eindrapport Masterplan Fiets. Samenvatting, evaluatie en overzicht van de projecten in het kader van het Masterplan Fiets, 1990–1997.* Den Haag, 1998.

Westermann, Andrea. *Plastik und politische Kultur in Westdeutschland*. Zurich, 2007.
———. "When Consumer Citizens Spoke Up: West Germany's Early Dealings with Plastic Waste." *Journal of Contemporary European History* 22, no. 3 (2013): 477–498.
Whitt, Frank R., and David G. Wilson. *Bicycling Science: Ergonomics and Mechanics*. Cambridge, 1974.
Wildt, Michael. *Am Beginn der Konsumgesellschaft. Mangelerfahrung, Lebenshaltung, Wohlstandshoffnung in Westdeutschland in den fünfziger Jahren*. Hamburg, 1994.
Wilson, David G. "The future of the bicycle and its potentialities." *Long Range Planning* 8, no. 2 (April 1975): 81–87.
———. "The future potential for muscle power." In *Pedal Power in Work, Leisure, and Transportation*, ed. James C. McCullagh, 106–124. Emmaus, PA, 1977.
———. "Getting in gear: Human-powered transportation." *Technology Review* 81 (1979): 42–54.
———. "The development of modern recumbent bicycles." In *Human-Powered Vehicles*, ed. Allan V. Abbott and David G. Wilson, 113–127. Champaign, IL, 1995.
Wilson, David G., and Jim Papadopoulos. *Bicycling Science*. Cambridge, 2004.
Wilson, Stuart S. "Bicycle technology." *Scientific American* (March 1973): 81–91.
Windmüller, Sonja. *Die Kehrseite der Dinge. Müll, Abfall und Wegwerfen als kulturwissenschaftliches Phänomen*. Münster, 2004.
Winner, Langdon. *The Whale and the Reactor: A Search for Limits in an Age of High Technology*. Chicago, IL, 1986.
Woodward, Donald. "'Swords into Ploughshares': Recycling in Pre-Industrial England." *The Economic History Review* 38, no. 2 (1985): 175–191.
Worster, Don. *The Wealth of Nature: Environmental History and the Ecological Imagination*. New York, 1993.
Wray, J. Harry. *Pedal Power: The Quiet Rise of the Bicycle in American Public Life*. Boulder, CO, and London, 2008.
Zelko, Frank. *"Make it a Green Peace": The Rise of Countercultural Environmentalism*. New York, 2013.
Zeller, Thomas. "Building and Rebuilding the Landscape of the Autobahn 1930–70." In *The World Beyond the Windshield: Roads and Landscapes in the United States and Europe*, ed. Christof Mauch and Thomas Zeller, 125–142. Athens, GA, 2008.
Zierenberg, Malte. *Stadt der Schieber. Der Berliner Schwarzmarkt 1939–1950*. Göttingen, 2008.

Index

1950s syndrome, 161, 171
2002 Johannesburg Summit on Sustainable Development, 131n33

A
Abfallwirtschaftsprogramm, 173
Accident, 21, 78, 92, 113
Accolatse, Alex, 18–19; and photography, 19
Accra, 2
Adelmann, Georg, 154
af Klercker, Carl-Henrik, 105, 107
Africa, 2, 8, 15–20, 22, 24, 27–28, 136, 157, 207, 222. *See also* West Africa
Aircraft, 65
Ajka, 185–190, 195
Ajka Alumina Factory (*Ajkai Timföldgyár*), 186, 189
Albert de la Bruhèze, Adri A., 48
Amerika, 40
Amsterdam, 41, 84, 86, 158
Anthropocene, 8–9; and environmental history, 9; periodization of, 9n21
Antwerp, 76
ANWB (Dutch tourist and traffic organization), 47–48, 76, 79, 81
Appiah, Kwame Anthony, 16
Asahi, 130
Austria, 75–77, 84–85, 153–154; Anschluss regime, 85
Austrian Federal Environmental Protection Agency, 193
Automobile, 1, 46, 60–61, 65, 67, 70, 73, 78–81, 88, 93, 103–104, 107–111, 130–131, 138, 217, 221–222
Automotive lobby: lobbying efforts, 79–80; members of, ANWB, 79; Dutch Basalt Company, 79; Dutch Concrete Society, 79; NV Bitumenweg, 79; Royal Dutch Automobile Club KNAC, 79; Union for Business Car Owners BBN, 79

B
Bakonyi, Árpád, 186, 190, 194
Ballantine, Richard, 38, 40, 43, 46; *Richard's Bicycle Book*, 40
Baltic Sea, 101
Barcelona, 90–91
Baschet, René, 61n12
Basel Convention, 190
Bayer process, 184
Beck, Kurt, 21
Beek, Guss van de, 45
Belgium, 41, 75–76, 89, 203
Berglund, Helge, 108, 113
Berlin, 203–210
Berliner Stadtreinigung (BSR), 203–205, 208–210
Bicycle, 1, 5, 7, 9, 15–28, 33–34, 36–51, 58–70, 73–93, 101–105, 107–116, 129–139, 215–217, 221–222
Bicycle Commuter Act (US, 2008), 89
Bicycle Information Point (FIP), 88
Bicycle taxes, 7, 73, 75–78, 81–82, 84–88, 91–92; and clubs in Europe, 76; and class politics, 77–79, 83–87, 91–93; and cycling advocates, 76, 86, 92; and European bicycling industry, 89; financial burden to workers, 84; and financing infrastructure, 74–75, 78, 85; and highways, 78–80, 84, 86; and history of, 74–77; and incentives, 88–89; in Liechtenstein, 92; in Netherlands, 79–81, 85, 88–89; and politics, 81, 86,

90, 93: and sustainability, 75, 86, 93; in Switzerland, 92; in the United States, 77, 86–87; and working class resistance, 84
Bicycle Thief (film), 73–74, 84. *See* Vittoria De Sica
Bicycle Transportation, 40
Bicycle Transportation Alliance (BTA), 91
Bicycles and West Africa: appearance, 21, 23–24; advocacy organizations, 33; obtaining spare parts, 22–23; as African modernity, 16, 18, 222; Ballantine's Design, 48–49; as commodities, 16, 20, 26; cultural appropriation of, 16–18, 20; everyday culture, 17–20; and gender, 26; history, 17–19; as investment, 15, 20, 25; mobility, 19–20, 22; new, 24, 28; objectification, 24–25; Raleigh Roadster, 19; recumbent, 37–38, 41–43, 49; repairs, 21; transport, 19–20, 22–23, 25, 27
Bike 2000 Construction Contest, 44
Black market, 65
Bonardi, Italo, 79
Bonn, 91
Borneo, 216
Bourdais, Roland, 68
Bowden Gregory, H, 19; on bicycle History in West Africa, 19
Brand, Stewart, 2, 5, 34, 39; *Whole Earth Catalogue: Access to Tools*, 2, 34, 39
Braungart, Michael, 10
Brownell, Emily, 195n32, 196
Brundtland, Gro Harlem, 4
Brundtland Report, 215. *See* World Commission on Environment and Development
Bücher, Karl, 14792
Burkina Faso, 15, 17, 20, 22, 25, 27–28
Burrows, Mike, 44
Bus, 61, 82, 107, 111, 132

C
Cairo, 2
California, 2, 40
Capitalism, 3, 216–217, 221
Car. *See* automobile

Carlberg, Charles, 109, 113
Carlowitz, Hans Carl, 3–4, 28; coining of the term Nachhaltigkeit or sustainability, 3; in connection to the *Report to the Club of Rome 2052*, 2n7; *Sylvicultura oeconomica or the Instructions for the Wild Tree Cultivation*, 3
Centralbron (bridge) 103, 105–106, 109, 112–113, 116; and bicycle traffic, 107; and car traffic, 105–109, 112–113; construction of, 105; and cyclists' federation, 107, 109; exclusion of bicyclists, 105–109, 112–114, 116; infrastructure of, 105, 116; and traffic patterns, 105–106; and views on by af Klercker, 105, 107
China, 19, 28, 127, 131, 202, 208, 216
Chinese Phoenix, 19
Chūō kōron, 130
Clear Channel, 90–91
Clothing, 25, 62, 126–127, 149, 154, 156–157; costs in Salzburg, 156
Clothing repair, 149–153; in 18th century Europe, 148–159; and associations with poverty, 154–155; costs associated with, 149–150, 154; and gender, 150; as household strategy; 149–152; as occupation, 149–153; and Romani people, 153–154; and taxes, 154–155; and urban Jews, 153; in Vienna, 151–153
Club des Villes Cyclables, 87, 89
Consumer, 1–2, 5, 7, 9, 25, 45, 127, 147, 150, 160, 192, 205; behavior, 177; demand, 218; goods 15, 17, 20, 25, 45, 66, 75, 196; society, 94, 169, 172–173, 216–217
Consumerism, 104
Consumers Guide to the Protection of the Environment, 2
Consumption, 20, 24–25, 28, 104, 128, 147–148, 155–162, 169, 171, 174–175, 178, 188, 197, 215–217
Copenhagen, 87
Corruption, 190–191

Coulon, Josepha, 68
Counterculture, 34, 46, 86–87
Cox, Peter, *Moving People*, 16
Crepeau, Michel, 90
Critical Mass (cyclists' interest organization), 33
Cronon, William, 224
Crutzen Paul J., 7, 219
Cycle Vision, 42
Cycling, 1, 5–8, 16, 19, 33–34, 36–38, 40–41, 43–51, 62, 73–79, 86–93, 101–103, 105, 107, 109–114, 116
Cycling Civil Servants Naaldwijk (FAN), 88; merge with non-profit foundation "Cycling to work? Do!", 88
Cyclists' Touring Club, 48

D

Deforges, Regine, 66; *Blue Bicycle*
Denmark, 41, 131, 168, 203
De Sica, Vittoria, 73–74, 84, 86; *Bicycle Thief* (*Ladri di Bicicletta*), 1948 (film) *Signal*, 63
Digital waste. *See* E-waste
Duschill, Josef, 85
Dutch, 46–47; bicycle design, 49; cycling culture, 46–48, 51; national identity, 47; recumbent bicycles, 50; recumbent bicycle firms, 43
Dutch Cycling Union. *See* Fietsersbond
Dutch Concrete Society, 79
Dutch Ministry of Transport, 88
Dutch National Road Fund, 81

E

East-Flanders, 76
Ebert, Anne-Katrin, 86
Eco-Chari, 136
Ecology, 4, 38, 195, 203, 212, 219–220
Edgerton, David, 5n13
Edo Period, 125–126, 128; and city of Edo, 126, 216; and government policy, 126; ideology during, 138; and population growth, 126; and recycling, 125–126, 138; and waste management, 125–126
Eisuke, Ishikawa, 128n11, 128n12

Electrical and Electronic Equipment Act or ElektroG (Germany, 2005), 204
Elektro-Altgeräteregister (EAR), 205
Engineer, 34, 39, 59, 75, 78, 81, 85, 101, 103, 105, 113–114, 171, 189, 204, 225
Engineering, 21, 24, 33, 36–37, 52–55, 73, 78–78, 81, 87, 189
Environmental Handbook, 2
Environmental history, 5, 8–9, 179, 224
Erlangen, 87
Ernfelt, Carl, 109, 114
Essingeleden, 105–109, 112, 114, 116; and bicycle lanes, 108; and car traffic, 107–108; and exclusion of cycling lanes, 105–108; and national road administration, 108, 114; and Gröndalsbron road section, 108, 114; as part of 1960 highway, 108
E.T. the Extra Terrestrial (1982) film, 46
Europe, 2, 8, 17, 28, 36, 41, 58, 60, 86, 90, 104, 129, 132, 147, 149, 153, 158, 189, 195, 204, 208–209
European Union, 8, 89, 189
European Union Waste Code, 189
European Waste Electrical and Electronic Equipment Directive (WEEE Directive), 204n9
E-waste, 202–210; in Berlin, 203–205, 209–210; as commodities, 203; and copper, 208; and costs of collection, 209–210; dumps in less developed nations, 202; and German media, 202; and global division of labor, 203–204; and manufacturing, 209; processing in Africa, 208; and net worth of industry; 208; and recycling of, 206–207; as scrap, 205, 207; toxicity, 202; and waste chain networks, 204–206; and waste handlers, 207; and waste scavengers, 208

F

Fiets, 41–43, 45
Fietsersbond, 44–45, 47
First World War, 19, 77, 80, 169
Fisch, Rudolf, 18
Florida, 75

Forester, John, 40
Forestry, 4
France, 41, 58–64, 66–69, 75–77, 87, 89, 157, 184; and French army's defeat in Second World War, 65; mobility culture, 58; negative views on biking, 63–65; and post WWII transport, 66–70; poverty and bicycle status, 66–67; Resistance, 63–66; under German occupation during WWII, 59–65; urbanization, 59; Vichy bicycle propaganda, 62–63
Freiberg, 3
French Resistance. *See* France, Resistance

G
German Bicycle Club (ADFC), 92
Germany, 3, 75, 79, 87, 91, 131, 168–169, 171, 174, 178, 184, 202–205, 207, 210–211, 216
Ghana, 2, 21, 27, 202–203, 208
Glass, 1, 63, 127, 170–174, 176–179, 205–206; and recycling 1, 168, 171–173
Global North, 196
Global South, 196–197
Globalization, 9, 28, 198
Gold Coast, 18
Goudig, Mam', 67
Grande Illusion (1937) film, 64
Grande Vadrouille (1966) film, 64, 72n22
Graz, 76, 85, 87
Graz University of Technology, 85
Great Acceleration, 9
Great Britain, 9, 19, 76
Green Party, 113
Greenpeace, 5, 193, 209
Green Transport Plan (UK, 1999), 89
Grellmann, Heinrich-Moritz-Gottlieb, 153; *Versuch über die Zigeuner* (Essay on the Gypsies), 153
Grober, Ulrich, 4
Gröndalsbron. *See* Essingeleden
Grüner Punkt, 176
Guérin, Françoise, 67
Guinness Book of World Records, 37
Gyurcsány, Ferenc, 190

H
Haeckel, Ernst, 4, 220, 224; and Ökologie, 4
Hainaut, 76
Hanley, Susan, 126n5
Henneking, Carl, 75
Hereu, Jordi, 91
Hetherington, Kevin, 197n34, 198
Highway, 78–80, 83–84, 86, 108–110. *See* motorway
Hitler Youth, 169
Ho Chi Minh Trail bicycle, 46
Hokusai, 128
Hölzl, Richard, 4
Hollandrad, 51
Holocene, 9
Horse, 61
Human Power, 37, 39
Hungarian Academy of Science, 193
Hungarian Aluminum Industry, 185–187; and privatization, 186–187
Hungarian Socialist Party, 190
Hungary, 183–186, 188–191, 194, 197; and the red mud disaster (2010), 184. *See* Red Mud
Hungary National Development Agency (Nemzeti Fejlesztési Ügynokség), 195
Hungary Waste and Secondary Raw Material Management Program, 194
Huré, Maxime, 90

I
Illés, Zoltán, 192
India, 19, 198, 208
Industrial Revolution, 9
Industrialization, 9, 126, 161, 223
Innovation, 1, 5, 9, 16, 27, 34, 38–39, 43, 49, 93, 194–195
Innovations studies, 4–5
International Human Powered Vehicle Association (IHPVA), 34–39, 44, 49–50; founding of, 34; in the United States, 36; in the Netherlands, 36; and bicycle innovation, 36, 43
International Human Powered Vehicle Movements (IHPVS): in America, 36–39; in Netherlands, 36; contests,

43–44; initiatives, 44; differences between American and Dutch HPV clubs, 43–44, 50; *HPV Nieuws,* 50
International Human Speed Championships (1975), 37, 38, 40
International Road Congress (1926), 78
International Subcommission on Quaternary Stratigraphy, 8
Iraq War, 38
Iron Curtain, 40–41
Italy, 73, 75–76, 79, 84–85, 153
IT-Altgeräte-Verordnung (ITV), 204

J
Japan, 40, 125–132, 134, 136–138; 2010 White Paper, 128; and bicycle abandonment, 131–136; and bicycle design, 138; bicycles and gender, 130; bicycle ownership, 129–131; and bicycle recycling, 136–138; and history of bicycle ridership, 129–131; and bicycle pollution; 138; Bicycle Usage Promotion Study Group, 137; and history of recycling, 125–128; and Minister of Commerce and Industry; 130; and Minister of Environment, 125; and the Society for the Advancement of Bikecology, 131; and Japan Bicycle Promotion Institute, 137
Japanese Legislators' Association for Promoting Bicycle Use, 131
JCDecaux, 90–91
Johannesburg Summit on Sustainable Development (2002), 131
Jour de fête (1949) film, 66n31
Jugendverband KJV, 85

K
Kalderásch, 153
Kamaishi City, 136
Kirk, Andrew, 34n9
Knittelfeld, 85
Knowledge, 59
Köster, Roman, 179, 223
Kollo, 20
Kornai, János, 192n21

Kovács, László, 89
Kyle, Chester, 36–38, 40, 43

L
Lake Balaton, 185
Lake Mälaren, 101, 108
La Rochelle, 90
Law for Promotion of Sorted Collection of Recycling of Containers and Packaging, 127
Leipzig, 91
Liège, 76
Ligue Nationale pour l'ameloriation des Routes, 76
L'Illustration, 61
Ljung, Harry, 107
Lomé, 18
London, 160
Longhurst, James, 75, 77
Louisiana, 63
Lübeck, 75
Lukes, Steven, 102
Lyon, 65, 87, 90

M
Magdeburg, 75
Magyar Alumínium Termelő és Kereskedelmi Zrt (MAL), 183, 186–189, 192–194, 195–198; and acquiring of bauxite and alumina plants for privatization, 186–187; and environmental remediation, and hazardous waste, 190; and mining minerals, 194; parliamentary investigation of, 191, 193, 198; and red mud disaster, 183, 188; and wealth, 191–192; and renting of disposal sites, 193; and waste reuse, 196–197
Mahoney, James, 102
Make (magazine), 2
Maryland, 75
Massif Central, 64–65
Maszobal, 186
May, Ernest, 224
McDonough, Michael, 10
McNeill, John, 219

Meadows, Dennis L., 4
Melville, Samuel Joseph, 46n79
Memory, 59–60, 63, 66, 68–69
Meyer, Birgit, 21
Migration, 20, 25
Milan, 78
Minneapolis, 75
Minnesota, 75
MIT, 37
Mitaka, 136
Miyata, 129
Moritz, Cor, 42
Motorcycle, 25, 67–68; and moped, 81
Motorway, 81, 109. See highway
Movement, 34
Mozart family, 149
Müll-Lawine, 173
Münster, 91
Municipal Coordinating Committee for Overseas Bicycle Assistance (MCCOBA), 136. See Japan
Musashino City, 136

N
Nairobi, 2
National Geographic, 40
National highway experiments, 79; and exclusion of bicyclists, 79
Nationalsozialistische Volkswohlfahrt (NSV), 169
Nazis, 85, 170
Nazi Germany, 169; and destruction of infrastructure, 170; organizations of, 169; and recycling, 169
Neolithic Revolution, 9
Netherlands, 36, 41, 43–48, 50–51, 65, 75–77, 79, 84–85, 87, 92, 131, 203
Neustadt, Richard, 224
New York, 75
Nieuws, 43–44
Niger, 25
Nigeria, 19–22
Nile, 21
Normandy, 66
Nuclear power plant, 1
Nuremberg, 150, 158

O
Obama, Barack, 89
Ökopol, 204
Ohio, 75
Oldenziel, Ruth, 48, 126
Omni, 37
Only True Bicycle Union (ENFB), 87–88. See also Dutch Cycling Union
Open Source Ecology Network, 3

P
Palmgren, Sture, 111
Paper, 20, 126, 167, 174–179, 215–216; as waste, 168–170, 174–179; and recycling, 171, 175
Paris, 61–63, 66, 70, 87, 90–91, 157–158
Pedestrian, 76, 78–79, 81, 85, 108, 112–114, 132, 134, 217
Peterson, Cary, 41
Peugeot, 67
Pfister, Christian, 147, 161n94, 171
Plastic, 16, 23, 127, 168, 170, 172–174, 177–179, 205–206
Pollution, 27, 38, 109, 111, 127, 131, 134, 136, 138–140, 177, 197
Population growth, 137
Portland, 75, 87, 91
Postma, Martine, 2; and Repair Café Foundation, 2
Privatization, 186–187, 191–192, 198
Proßnitz, 149, 153
Provo movement, 86–87
Püttlingen, 3

R
Rail, railroad, railway. See train
Ramazzini, Bernardino, 153
Raymond, Charles T., 75
Recycling, 1–2, 5–8, 10, 125–129, 136–138, 147–148, 161, 168–171, 173–179, 183–184, 194–197, 201–211, 215–217, 219, 221–223
Red Mud, 183–195, 197–198; and alumina industry, 185–186; chemical composition of, 184; classification as hazardous waste, 190; disposal

methods, 184–185; and holding ponds, 187–188; and hazardous waste policy, 188; industrial history, 185–186; privatization of lagoons, 186–187; and pond leakage, 188–189; for-profit uses of, 191–195; and recovery of minerals, 194; risks of disposal, 185; toxicity, 185; toxic sludge disaster, 183

Rennes, 87, 90

Repair, 2, 21–22, 24, 26, 28, 87, 130, 136, 147–155, 157, 159–161, 170, 223; shops, 19

Repair trade, 147–161

Repair Café Foundation, 2

Resell, 147–148

Restriction of Hazardous Substances Directive (RoHS Directive), 204n10

Reststoff-Bearbeitungs GmbH (BRAL), 205, 209

Reuse, 93, 125–126, 128, 135–136, 148, 155, 160–161, 171–172, 184, 194–198, 223

Rickshaw, 61–62, 129–130, 139

Rio de Janeiro, 3

Rochester, 75

Romani. *See* Clothing repair

Rubenstein, Bob, 41

S

Sabin, Paul, 224; and "The Ultimate Environmental Dilemma" (2010), 224

Sachs, Wolfgang, 110

Salzburg, 149, 152, 156

San Francisco, 2

Schenau, Van Ingen, 43

Schiffmann, Traugott, 85

Schumacher, Ernst Friedrich, 2; *Small is Beautiful: A Study of Economics As If People Mattered*

Schomburgk, Hans, 18

Scientific American, 33–34, 40

Scott, James C, 4n9; *Study on Modern Statecraft*

Secondhand trade, 6, 147–148, 153, 159–161, 202, 207; and consumption, 147–148, 155–157, 160; and recycling, 148, 151; and reduction of expenses, 148, 155–157; and sustainability, 147–148, 150, 160; and taxation of traders, 156–158; and trade, 155–160; in Vienna, 151–152, 155–157; and urban Jews, 159–160; and women, 159

Second World War (World War II), 6, 19, 37, 59, 85, 103–104, 127, 161, 168–170, 173–174, 179, 216, 222

Shanghai, 19

Shigeki, Kobayashi, 137

Shozo, Tanaka, 127

Shove, Elisabeth, 6, 9, 102–103; on cycling history, 6; on linear history, 6

Šiauliai, 41

Sijbrandij, Ymte, 50n95

Small Earth (*De Kleine Aarde*), 43

Smit-Kroes, Nelie, 88; and Dutch Ministry of Transport

Smithsonian National Museum of American History, 46

Society for the Advancement of Bikecology (Japan), 131

Soviet bloc, 188

Soviet Union, 186

Spängler, Franz Anton, 149

Spielberg, Steven, 46; *E.T. the Extra Terrestrial* (1982) film, 46

Spittler, Gerd, 25

Stadtguardia, 151

Städtehygiene, 171

Städtetag, 171

Steele, William M., 216

Steffen, Paul, 219

Steyr-Daimer-Puch AG, 85

Stockholm, 4, 101, 103–107, 109–112, 114, 116; and 1960s attitudes towards bicycles, 109–112; and Alternativ Stad, 110n3; and rise of car ownership, 103; and social status of cyclists, 111; and the urban environmental movement, 111, 116

Stockholm Chamber of Commerce, 112

Stockholm Central Board of Administration, 105

Stockholm Master Plan (1952), 107n23

Stoermer, Eugene F., 8
Stöger, Georg, 223
Strandberg, Samuel, 101
Strasser, Susan, 147; *Waste and Want*, 147
St. Paul, 75
Strasbourg, 87
Stuttgart, 92
Styria, 84–85
Sudan, 21
Sustainability, 1, 3–6, 8, 10, 27–28, 74–75, 86, 89, 91–93, 125–126, 128, 147–148, 150, 168, 183, 215–216, 218–219, 220, 222–224; goals, in German forestry production; 3–4; as commodity, 4; and history, 3–10, 126, 215–216, 218–220; by recycling, 128, 147; by bicycling, 27–28; 86, 89–93; in the chemical industry, 183; and second hand trade, 42, 67, 84, 148, 151, 153–159; and household waste in West-Germany, 168
Sweden, 87, 103–104, 107, 131; and bicycling, 103–104; car ownership after WWII, 103–104; cyclists federation, 107–109; national road administration, 108, 114; urban planning, 104–105
Switzerland, 19, 41, 75–76, 92, 168
Szirmai, Viktória, 188–189, 194

T

Tati, Jacques, *Jour de fête* (film) 1966
Tax, 4, 7, 15, 19, 32, 38, 48, 66, 73–102; taxation 74, 76
Technical University Eindhoven, 44
Technology, 1–3, 5–10, 16–17, 21, 28, 33–34, 36, 38–39, 41–43, 60, 67, 69, 93, 125, 129, 138, 177, 183–184, 187, 192, 194–198, 207, 2016; alternative, 24, 43
Technology users, 5–6
The Hague, 41, 81, 83
Tiébélé, 20
Tierney, Thomas F., 221
Tijken, R., 81
Togo, 23
Tokyo, 127, 134–136; and bicycle commuters, 134; and local governments, 134; population of, 134; and problems with abandoned bicycles, 134; taxes, 136
Toshima, 136
Traffic count, 80–81, 83
Train, 61, 68, 102, 112–114, 134, 136
Train station, 132, 136
Tram, 61, 107
Trischler, Helmuth, 126

U

UN Conference on the Human Environment, 4
UN Food and Agriculture Organization (FAO) (1951); *Principles of Forest Policy*, 4
UNESCO's Man and Biosphere Program (1970), 4
United Africa Company, 19
United Nations' Brundtland Commission (1987), 4
United Nations Environment Program (UNEP), 202
United Nations University (UNU), 202
United States, 2, 36, 38–40, 45–47, 51, 75, 77, 79, 86, 89, 93, 129, 131, 179, 208, 216, 222; and bicycling infrastructure, 45n75; Congress, 89; cycling culture, 45–46; E.T. film and bicycle identity, 46; national identity, 46
University of Oxford, 33, 37
Urban Jews. *See* Secondhand markets
Urban planning, 8, 73, 104–118, 178
User. *See* Consumer

V

Van de Beek, Guus, 45
Vélosolex, 67–68; alternative to car, 67; and gender, 68; as social indicator, 68; and teenagers, 68
Verrips, Jojada, 21
Verpackungsverordnung, 177
Vienna, 151–155, 157–159

W

Wasserhaushaltsgesetz, 173

Waste, 1, 5–6, 26, 125–129, 136–137, 148, 168–184, 188–191, 193–198, 102–216
Waste Dependent Development, 195
Waste management, 93, 169, 171, 173, 176–177, 179, 194, 203–204, 207
Waterlooplein, Het, 84
West Africa, 15–28; and bicycling. *See* Bicycles in West Africa; Burkina Faso, 22, 25, 27
Westernization, 129
West-Flanders, 76
West Germany, 6, 41, 168–172, 174–175, 177–179; and consumer attitudes towards recycling, 177; as economic miracle, 168, 170; and glass, 168, 170–174, 176–179; and household waste, 171, 176, 183; and housewives' changing behavior, 170; and plastics, 168, 170, 172–174, 177, 179, and plastics industry, 172; and waste, 168–179; and waste composition, 170; and waste disposal, 168, 173; and waste paper, 174–176; and privatization of recycling, 169, 176, 179; and recycling, 168–171, 173–179; as recycling role model, 179; and revival of recycling, 173–176
Willkie, Fred, 38
Wilson, David Gordon, 37–38, 40, 43, 45; *Bicycling Science*
Wilson, Stuart, S., 33–34, 37
Windmill, 1
World Bank, 17
World Commission on Environment and Development, 215; *Our Common Future*
World Naked Bike Ride, 33
Worster, Donald, 10, 220, 223

Y
Yamaguchi Bicycle, 130
Yumenoshima, 127

Z
Zoo, 18
Zucca, André, 63

www.ingramcontent.com/pod-product-compliance
Lightning Source LLC
Chambersburg PA
CBHW072150100526
44589CB00015B/2160